FORMS, TEXTURES, IMAGES

a photo-essay by Takeji Iwamiya

edited, with an introduction, by Mitsukuni Yoshida

with an appreciation by Richard L. Gage

FORMS, TEX

TURES, IMAGES

Traditional Japanese Craftsmanship in Everyday Life

WEATHERHILL/TANKOSHA

New York · Tokyo · Kyoto

This book was originally published in Japanese in 1978 by Tankosha, Kyoto, under the title *Nihon no Katachi* (Japanese Forms). Mr. Yoshida's introduction has been translated by Susan Carol Barberi and Mr. Iwamiya's photographic commentaries by Richard L. Gage. The quotation on page 7 is from *The Unknown Craftsman: A Japanese Insight into Beauty,* by Sōetsu Yanagi, founder of the Japanese folkcraft movement, translated and adapted by Bernard Leach and published in 1972 by Kodansha International.

First English edition, 1979

Published jointly by John Weatherhill, Inc., of New York and Tokyo, with editorial offices at 7–6–13 Roppongi, Minato-ku, Tokyo 106, Japan, and Tankosha, Kyoto. Printed in Japan.

Library of Congress Cataloging in Publication Data: Forms, textures, images. / Translation of Nihon no katachi. / 1. Handicraft—Japan. / 2. Design, Industrial—Japan. / I. Iwamiya, Takeji, 1920–/ II. Yoshida, Mitsukuni, 1921–/ III. Gage, Richard L. / TT105.N4913 / 745'.0952 / 79–15714 / ISBN 0–8348–1519–2

Contents

The special quality of beauty in crafts is that it is a beauty of intimacy. Since the articles are to be lived with every day, this quality of intimacy is a natural requirement. Such beauty establishes a world of grace and feeling. It is significant that in speaking of craft objects, people use terms such as savour and style. The beauty of such objects is not so much the noble, the huge, or the lofty as a beauty of the warm and familiar. Here one may detect a striking difference between the crafts and the arts. People hang their pictures high up on walls, but they place their objects for everyday use close to them and take them in their hands.

—Sōetsu Yanagi

Aspects of Form in Japan: An Introduction

by Mitsukuni Yoshida

FORM AND COLOR The Roman architect Vitruvius presented his great work on architecture to Augustus Caesar sometime in 25 B.C. This work, divided into ten books, exerted a powerful influence on European architecture that continues to this day. In a description of the ideal architect of the Augustan age, Vitruvius stated that he should have an understanding of literature, should excel in geometry, be well versed in history and philosophy, and also have adequate knowledge of music, medicine, law, and astronomy. It was necessary for an architect to be a man of many talents.

What is most interesting here is the importance Vitruvius placed on geometry. Since Greek and Roman times the principles of geometry have been considered basic to a wide variety of subjects. Vitruvius laid down the principles of geometric proportions and ratios as the foundation of architectural design. In his day it was already commonly accepted that the ideally proportioned human figure was eight heads tall. Vitruvius went further in his definition of perfect beauty and applied this ratio to all parts of the human anatomy. Accordingly the ideal human figure, with extremities outstretched, described a circle with its center at the navel, and when lines were drawn between the tips of the fingers and toes of the extended limbs, a perfect square resulted. In this manner Vitruvius used the human body itself to demonstrate the basic principles of geometry. It was only natural, then, that architectural forms, such as temples, should be designed to create geometrically proportioned spaces, for the perfectly proportioned human figure would first appear in its greatest beauty when placed within the context of a geometrically ordered space.

Vitruvius designed not only temples but also theaters, government buildings, and private residences. He detailed the proper proportions of Doric, Ionic, and Corinthian columns and explained how they should be drafted on paper. All could be reproduced with the aid of ruler and compass. Geometric proportion was the fundamental principle whereby all spatial and physical forms, including the human body, could be determined.

This ordering of the world in geometric terms was even further clarified by such men as Plato, who used geometry to explain the creation of an orderly universe out of chaos. Plato based his theory upon two types of right-angled triangles: one formed by cutting a square diagonally in half, and the other by dividing a regular triangle into two parts. Various combinations of these triangles created the particles of matter that in turn formed the four elements of the universe. Thus a regular tetrahedron (four-sided solid) represented fire, a regular octahedron (eight-sided solid) represented air, a regular icosahedron (twenty-sided solid) represented water, and a regular hexahedron (six-sided solid) represented earth. A fifth element, the heavens, was composed of particles of icosahedrons that were formed by regular pentagons. The whole universe was in the shape of a ball, and every portion of this ball equaled every other portion, for the whole had a perfectly symmetrical shape. The circular motion of the universe was natural because this was a perfect geometric form. A rotating universe within which globular heavenly bodies described circular orbits was a firmly established concept in European astronomy. Even Copernicus, who greatly enlarged man's conception of the universe by replacing Ptolemaic theory with his heliocentric theory, was not completely free from the spell of the Platonic tradition. The Copernican theory still rested upon the premise that all heavenly bodies were round and had circular orbits.

In these various ways geometry was employed in ancient days as a device to fix and define forms; it was a means of conceptualization. It was for this reason that the Greeks fashioned the statues of their gods with the ideal proportions defined by geometry. A god was a creature having ultimate beauty, and for the Greeks the ultimate ideal of beauty was always to be found in the perfect human figure.

China, ancestor of much of Japan's cultural heritage, also used geometry as a basic principle of thought. There is a Confucian classic, the *Chou Li* (周礼), that outlines the ideal organization for the government of the ancient Chinese kingdom of Chou. Proper modes of human behavior were defined in terms of *li* (礼), or etiquette. According to this system, the bureaucracy was divided into six ranks of heaven, earth, spring, summer, fall, and winter, each rank having its own distinctive cap. The *Chou Li* painted a picture of a utopia ruled by a systematically organized bureaucratic monarchy.

The chapter of the *Chou Li* describing the sixth rank of winter has been lost, but in its stead another chapter, entitled "K'ao Kung Chi" (考工記), has been handed down. This chapter describes in detail the standards of production for a variety of arts beginning with architecture. The "K'ao Kung Chi" ranks with the writings of Vitruvius and serves as a most important source of information on ancient Eastern technological philosophy.

Within this text there is an explanation of how to derive the measurements of an object according to the ratio of its proportions. The basic unit was set at 8 *ch'ih* (in Japanese, *shaku*), the average height of a man in those times. One *ch'ih* was equal in those days to approximately 21 centimeters, making the standard measure 168 centimeters. Multiplying 8 *ch'ih* by one-third and adding to this one-half of 8 *ch'ih* produced the figure of 6 *ch'ih*, 6 *ts'un* (in Japanese, *sun;* 10 *ts'un* = 1 *ch'ih*). This was the diameter of a horse-cart wheel. In another use of the human figure as a standard of measure, the head was said to equal one-sixth of the total height, the ideal human height therefore being six heads. This method of using ratios as standards of measurement resembles that employed by Vitruvius, but the idea is not expanded to use geometric forms for the construction of objects; the problem of measurement was only approached in straight mathematical terms using whole numbers.

Despite this, there are strict rules of geometric symmetry to be found—for example, in a city plan laid out according to the principles of *li* (礼). The ideal Chinese city was shaped in a square, each side measuring nine *li* (里). This square was divided by nine vertical and nine horizontal avenues intersecting at right angles. The palace stood in the center of the grid formed by the intersecting avenues. Each avenue had nine lanes. A wall surrounding the city had two gates on each side with three roads stretching out from each gate. This one example shows how *li* (礼) was used to define and organize all Chinese forms, for this city plan is based on such geometric forms as the nine vertical and nine horizontal lines and on the shape of a square.

Plato spoke of the world in geometric terms of polyhedrons, many-sided solids. In his philosophy it was necessary to be able to reduce all forms of existence to geometric terms. What, then, was used in China to define the world? One definition is that of yin and yang, often spoken of in terms of odd and even numerical values, both being types of dualism. There was also the theory of the five natural elements symbolized by the number five. Like the four elements of Greece, these represented the basic elements of nature: wood, fire, earth, metal, and water.

Originally considered to be the five necessary components for human existence, these five elements or substances were very real, tangible objects. Later, as they took on a more abstract character, they came to be known as the five principles. These five principles were thought to be interconnected, undergoing constant change and rotation.

The concept of the five principles gradually broadened to encompass the basic elements composing the world and the entire universe. The five principles have been represented in many forms. There are the five legendary virtuous emperors, the five directions (the four cardinal directions and the center), and the five seasons (spring, summer, fall, winter, and the season of transition between each of these). In human terms there are the five senses, the five internal organs, and the five tones that can be distinguished by the sense of hearing. All earthly organisms are classified into five types, and the five basic colors are blue, yellow, red, white, and black. Ethics are divided into the five virtues of benevolence, justice, courtesy, wisdom, and sincerity. The five principles penetrated every form of existence in the world. Geometry was not the key to the universe, and nowhere does it appear in these symbols.

Chinese descriptions of the movement of the planets and the stars were vague and cloaked in legend. There were three ancient theories concerning the universe: *kai t'ien* (蓋天), *hun t'ien* (渾天), and *hsüan yeh* (宣夜). In the *kai t'ien* view of the universe the earth was a flat plane upon which the heavens rested like an inverted bowl. The two shapes these represented, the circle and the square, were considered to be perfect forms. This theory also established a right-angled triangle, its sides having a ratio of three-four-five, as a standard of measure, but the use of geometric figures stops here. There was no further attempt at combining or developing these forms.

The universe of *hun t'ien* was even more conceptual. Heaven and earth were compared to a bird's egg, with heaven containing the earth as an eggshell contains the yolk. The metaphor was not elaborated upon, nor was the actual shape of the egg defined. According to the *hsüan yeh* theory, heaven was not an actual physical body but rather an infinite void. This automatically precluded any consideration of finite forms. The common bond that tied all existence together and corresponded to the five principles was not form but color.

This relation of the five principles to five colors rather than five forms is what most distinguishes Chinese culture. An object's being was symbolized by color, its existence clarified by color. In a broad sense this could be called a metaphysics, a phenomenology based on color. In the oldest Chinese dictionary, dating from Han times (206 B.C. to A.D. 220), Liu Hsi gave definitions for many Chinese characters. In the fourteenth chapter Liu explained the names of colors. "Blue (青) has the meaning of life (生); it is the color of creation. Red (赤) has the meaning of burning brightly; it is the color of the sun. Yellow (黄) has the meaning of sparkling light; it is the color of sunshine. White (白) has the meaning of opening or clearing; it is the color of cracked ice. Black (黒) has the meaning of darkness; it is the color of dawn and evening. Vermilion (絳) has the same sound as 工 (artisan's skill); it is a difficult color to dye and has the meaning of skill or craftsmanship. Purple (紫) has the same sound as 疵 (wound) and is not a true color; it is like a stain or flaw that deludes people," and so on.

Liu Hsi believed that colors captured the very essence of matter and all phenomena. Therefore color was a symbol of reality, and objects were to be distinguished by color rather than by form.

This mode of thought was transmitted to Japan, ever open to influence from China, and it soon became an accepted method of representing important objects. The result is clearly evident in Japan's many spiritual rituals and esoteric ceremonies. Rituals for the gods first took on meaning when conducted in an area decorated with the five colors. An object came into being when the duality of color and form were united. An example of this seen in a particularly unique Japanese form of beauty, the *kasane-no-irome* of the Heian court (794–1185). Another example is the splendid *odoshi* found in the armor of the samurai of a later period in Japanese history. The *kasane-no-irome* consisted of many layers of robes, each robe being a single color. It was the final combination, the relation of each color to the next and the overall harmony of the whole wherein the beauty lay. A similar effect was achieved by the many pieces of colored leather tied together with varicolored silken cords that made up the samurai armor. Eventually the same colors were used to distinguish classes and vocations in society. Ranking by color had already been in use in T'ang (618–907) and Sung (960–1126) China. Japan adopted the system in the Heian period with the establishment of forbidden colors. Every court rank and grade of office in the bureaucracy had its own color, no one being allowed to use colors reserved for a higher grade than his own.

The use of red and white, however, dates back to an earlier era before the introduction of the five-color system from China. These colors are what remains of inherent primitive Japanese religious beliefs. Red is the color of blood. It is the color of the life force, a symbol of life itself. Red also often symbolizes resurrection or rejuvenation. Red as a symbol of life is common to nearly all the ancient cultures of the world.

White was a symbol of all that was sacred. In ancient China white vessels, elaborately decorated, are thought to have been used solely as ceremonial wares. Confucianism designated white as the primary sacred color, just as black was made the sacred color of Hades, the color of death. White represented the highest level of godliness in this world, while black represented the unknown world on the other shore.

This tradition still exists today. Many objects in daily use in Japan are red and white, most of them for religious ceremonies. Wrapping paper and *mizuhiki,* a form of decoration made of twisted paper cords, for auspicious events are red and white or gold and silver. These symbols alone are sufficient to evoke a sense of occasion. Should the combination of colors be black and white, their meaning instantly changes, representing misfortune and unhappiness. The different ways of folding paper could also speak of changing fortunes, but the additional variation in colors, red and white or black and white (sometimes yellow), helps to complete the meaning. It is important to note here that this system of combining forms and colors to convey different meanings is one of the distinctive characteristics of Japanese forms. Representation of good fortune with gold and silver can be found in almost any culture. Gold is the color of the sun, silver of the moon, and together they rule the heavens. Both are also symbols of eternal unchanging time.

The Shinto shrines scattered throughout Japan are the sacred domains of Japan's type of nature worship. At the entrance to each and every one of these holy spaces stands a torii. Some are made of red-lacquered wood, some of stone. The symbolism behind red and white can be seen here too in the use of plain unfinished wood and red lacquer. As for stone, it is an unchanging thing symbolic of eternity. It was once thought that stone could propagate itself. In Western alchemy it was believed that precious metals such as gold and silver could reproduce themselves, and in the same sense in Japan sacred stones were thought to reproduce constantly in a demonstration of the everlasting nature of time.

FORM AND MATERIAL The stones that marked the boundary of a sacred enclosure were themselves thought to be immortal, and because of this they lent a sense of mystery to their forms, which declared the entrance into a closed world. One form was an adaptation of a gate marker to an enclosed village. The marker evoked an image of the unity of a village community, and there are still some villages in Asia that use this type of torii shape at their entrances.

Another object commonly made of stone is the *tōrō,* or stone lantern. Most of the *tōrō* found in Shinto shrines, Buddhist temples, and private gardens are made of stone. Why are they always of stone? Again, the reason is symbolic: to be found in the very nature of stone. Stone does not burn and is therefore a suitable container for fire. As was noted before, stone is also an everlasting thing. The flame glimmering inside the *tōrō* was a sacred fire, a symbol of life. In Europe the grace of God is often represented by light, and the advancement of civilization is commonly symbolized by light. The imagery of light bespoke the goodness of the gods and a flowering culture. Similarly in the East, light in opposition to dark has been symbolic of good versus evil. Light shining out of darkness, light brightening dark places, was a sign of victorious gods, of the ever-expanding power of the gods.

The fire offered at the Shinto and Buddhist altars in Japan has traditionally conveyed this sense of power. The divine flame proclaimed the might of the gods and symbolically sent forth this strength in the four cardinal directions. The *tōrō* was designed with this symbolism in mind. Its majestic appearance was an attempt to convey the majesty of the gods. Eternal light shining forth from a container made of the eternal substance, stone, helped to reinforce the sense of otherworldliness.

Gradually the *tōrō* came to serve as a source of light in gardens. The garden was originally a place set apart as sacred to the gods. Even today the Japanese garden is meant as an object to be viewed with a tranquil mind and not to be touched. In their search for a means of transcending the world, the Japanese created a garden that was a miniaturization of nature, a concentration of the elements of nature into one small space. Many such gardens had stones as their focal point. Occasionally there are combinations of three stones in a representation of the three Buddhas. This was an attempt to infuse stones of odd and unusual shapes with esoteric meaning. Legends of Japan often depict holy figures on rocks or stones, and many a stone is treasured because some deity is thought to have rested upon it. These are just some more examples of the use of stone to mark sacred boundaries. The flame within the stone *tōrō* seems a very natural component of the garden, centered as it is upon a stone or combination of stones.

As the sacred nature of the garden was forgotten and lost, the shape of the *tōrō* took on a more literal aspect, becoming a simple interest point within a carefully conceived landscape. In this case the *tōrō* also served to declare the presence of man, the reality of the human world, in a natural setting. The light within the *tōrō* no longer evoked images of some transcendent being, but began, rather, to encompass all that brightens and warms the world of man.

A stone torii, stone *tōrō,* and stone staircase all have in common the nature of the material of which they are constructed, its physical properties, and the additional characteristics that are attributed to it. In this case the material, stone, has given birth to the nature of the forms. The esoteric, magical qualities of the forms gradually evolved to become more symbolic of the secular world, of the things of man. This is how the characteristics of a certain type of material will determine the form into which it is shaped.

A similar limitation of the original material upon the final product is to be seen in the bamboo craft of Japan, southern China, and Southeast Asia. Bamboo is extremely resilient. This quality is not lost even when the bamboo is split, for it is an extremely tough material.

The vivid color and appearance of bamboo, its ability to grow as much as a meter in a night, combine to give bamboo its symbolic attributes. A well-known tale involving bamboo concerns the legendary seven hermits of the Six Dynasties period (222–589) in China. These sages renounced their worldly concerns to retire to a verdant and fresh bamboo grove where they whiled away their time in composing and reciting poems and exchanging drinks of wine as they conversed on pure and noble topics. They pursued the purity of a life of solitude removed from association with the vulgar world. A quiet life in distant mountains and solitary valleys where one might boil a cup of tea in lonely peace was the ultimate means of creating a sphere of tranquility. Thus did the seven hermits lead their pure lives in the bamboo grove.

In another Chinese legend one of the original virtuous emperors built a shrine dedicated to nature in his southern garden. A phoenix, bird of good omen, descended into the garden, bearing in its beak the seed from which the bamboo is said to have sprouted. The phoenix never alighted on the ground but perched in a paulownia tree. This legend is the source of the famous motif of a phoenix in a tree often found on imperial robes.

The abundance of such bamboo utensils as the tea scoop, water dipper, and tea whisk used in the uniquely Japanese art of the tea ceremony is another testimony of the purity attributed to bamboo. While bamboo on the one hand has proved useful in a very practical sense, it has also, on the other hand, helped to create an image of freshness and purity. In the *Ch'a Ching,* the T'ang-dynasty Chinese writer Lu Yü spoke of tea as a beverage that lent itself to a life of detachment. The Japanese tea ceremony, in its attempt to create a pure atmosphere on a high plane of consciousness, was a more elaborate means of attaining the same end. It has been to further this purpose that so many tea utensils have been made of bamboo.

The great tea master Sen Rikyū promoted the use of a cut bamboo section as a flower container for the tea ceremony. Until his time, flower containers were usually of bronze or porcelain. Rikyū saw the fresh and verdant bamboo as a more fitting material for conveying a sense of cleanliness and purity. His bamboo flower container was nothing more than a section of bamboo split in half, and yet this was a form that most strongly and cleverly revealed the physical characteristics of bamboo.

The resiliency and strength of bamboo made it a suitable material for the weaving of baskets. Today there are over twenty different patterns of weave for bamboo baskets. These include *yottsume, muttsume* (lattice weaves), *janome* (openwork resembling a snake's eye), a hexagonal pattern suggestive of a tortoise shell, a chrysanthemum design, the *tessen* (clematis) pattern, and matting—all with names that immediately convey their nature. All of these patterns of basket weave are of regular geometric shapes, surprisingly symmetrical for motifs of Japanese design. This might not be thought so odd, however, when it is considered that the same patterns are to be found in many other cultures, woven with the strongest and most resilient fiber to be found in each locality. In China the wisteria vine and in Southeast Asia varieties of palm are used. Each material is used in a manner best suited to its natural properties. The universal underlying principle of geometric patterns that results regardless of what material is used in the weaving helps to explain the symmetry of the Japanese baskets.

Another material commonly used to make containers in Japan is wood. In fact, wood has been used to such an extent that Japan is often known as the country of the wood culture. Wood is used in a variety of ways, but every unique form fully exploits the potential of the medium. One such method is to handle the material in a manner calculated to enhance the natural grain of the wood. Depending on how a piece of wood is cut, a straight or a cross grain can be revealed. Wood grain has always been important to Japanese woodwork. An object was designed so as to bring to life the natural lines of a straight grain or the cloudlike effect of a cross grain—another example of how form is regulated by the material used. Wood grain varies with every type of tree. The different forms to be found among objects made of wood are too numerous to list here.

Compared to such natural materials as stone, bamboo, and wood, forms created of metal and paper are the first representative types of man-made materials. From the beginning these materials have qualities created artificially by man, and therefore the forms made of them are not as limited. On the contrary, they can be said to have been created for the very purpose of going beyond the limitations imposed by nature. The history of the usage of such man-made materials is quite ancient.

The aforementioned classical Chinese technical guide, the "K'ao Kung Chi," has a section called "The Six Types of Metal." Bronze is an alloy derived by combining copper and tin. The quality of bronze varies according to the ratios of copper and tin. In "The Six Types of Metal" these ratios are discussed, giving six possible combinations, the objects to be made from each alloy, and the function of each object. Weapons such as halberds, arrowheads, and spears, as well as musical instruments, are categorized according to the different ratios of copper and tin. The ratios are determined by the function of the object to be made, the alloy being decided by the function. The quality of the material does not decree the type of form; the reverse is the case, the form being the prerequisite leading to the creation of the material. Bronze and other metal alloys are all made on this premise. Man's efforts are exerted toward finding the best techniques for creating forms from the alloys. Casting, wrought gold, and chasing are some of the means whereby forms of a conceptual nature are made concrete. Each form had to meet the requirements of the desired function.

Paper is much the same. The principle of papermaking, in which natural plant fibers are rearranged into a new flat form with the aid of an adhesive medium, allows for great flexibility according to the type of plant fiber, adhesive material, and technique used. The paper can be thick or thin and of varying strength. When a gloss is desired, material that will give the paper a patina is added. When a particular pattern is the aim, different fibers can be added and dyes applied. Many different techniques of making paper have evolved, and any number of different forms could be constructed out of paper.

The "K'ao Kung Chi" defines how an object is dependent upon the material it is made of and lists this principle under the heading of techniques. The same text says, "Time is in heaven, *ki* (atmosphere or energy) upon the earth, beauty lies in materials, and skill in craft; combining these four leads to good." The heavens represent the seasons; in other words, the appropriate season must be chosen. Earth—that is, all the natural conditions of the surrounding environment—must be satisfied. The materials in hand must be by far the most beautiful available, and only then is the craftsman free to apply his skillful techniques. When these four qualifications are met, then at last a product of quality is produced.

Theories of modern design carry this principle further in teaching that an object that best fulfills its function is automatically a form of beauty. When objects are made according to their materials, when a careful study has been made of the natural properties of the materials, then the form can emerge spontaneously. Only after these steps have been taken can the form be conceived—as if it only slept within the material, waiting to be released. Subtle signs indicating the presence of the final form can be read in the material while it is still in its natural state. The craftsman of old did not conceive of his form until after he had read these signs. His tools, extensions of his body, were then able to work as if with a life of their own. As the tools worked, the form emerged. At this point the character of the raw material no longer acted to limit the form but helped to build and fix it.

FORM FOR THE INDIVIDUAL AND THE COMMUNITY Man has long been known as *homo faber:* man as maker. Man strives to survive. In this effort he directs his body with his mind, at times conflicting with the nature around him, at other times moving in harmony with these forces. Man protects and preserves his life through the use of many devices and laws.

When these devices or tools were manufactured from materials found in nature, then each and every piece was something unique. Man adapted and used these forms to meet his life needs. A great variety of objects handcrafted from natural elements have come to surround man in his daily existence, and it is this fact that has earned him the name of *homo faber.*

Nature has infinite potential, but man has put to his use only one small part, the finite or limited aspect of nature, in his fight for self-preservation. Because his resources were limited, man did not dare to indulge in willful exploitation. Had he done so, he would have soon exhausted the vital supply of raw materials and found himself starving, facing death. Constantly threatened with this danger, man has had no choice but to learn, after much trial and error, to organize his technical knowledge into a rational system. The development of techniques enabled him to stretch and widen the range of nature available for his use. In the ceaseless quest for new raw materials, new techniques continually evolve, and each source of new material sets new limitations upon man.

Man has applied his talents to the materials found about him to create useful forms necessary to his existence; his success or failure in this undertaking relates directly to his success or failure in life. The creation of forms assures man security; they make his existence a more certain thing. Making forms is thus a means of self-recognition, a means of reassuring man of his place in nature.

A spear could be simply fashioned by carving a piece of wood. When a man used this spear to kill game, he assured his own continuation. As for the spear itself? In the beginning the end of a stick was scraped into a fine point. Later, through experimentation, it was found that grooves along the sides of the cone or triangle of the point increased its effectiveness. Once this was proved, men made grooves in their spearheads, and in this way a new form was disseminated among many people.

Improvement was constantly taking place. Every new design meant the invention of a new form. This did not mean, however, that every form underwent adaptation to suit it to man's evolving life styles. Man desires order and regularity in his life. He exists as a life form having a finite nature; it is impossible for an individual to undergo constant change. For this reason it was necessary to develop certain patterns or modes of life. The variety of forms necessary to each life system determined the quality of existence. In the beginning stages a form is chosen for its utility and then, later, for both its utility and its beauty. Security derived from a life style centered upon these forms. The continuity of a life system was the continuity of the human race itself. Each form has been a landmark, for in its particular time it expressed man's self-recognition. For the same reason it was also a guidepost to the future.

With the passage of time, forms were standardized and fixed into regular and constant shapes. If form is considered to be the declaration of the individual, then pattern is the confirmation of group unity. A group can have a spatial horizontal formation, as in the village community, or it can center around a time axis, as in an ethnic group. Whatever the type, patterns established within the group framework gradually become fixed, with the group as the basic unit. This standardization does not just take place in an ordinary daily continuum but in an extraordinary space-time continuum as well.

One example of this is seen in the changes that have taken place in the development of eating utensils. Eating is a direct maintenance of human life. For this purpose many ritualistic elements have been preserved to this day. Food vessels became quite prolific in Japan in the Yayoi period (200 B.C. to A.D. 250). The famous excavation sites of Karako in Nara Prefecture and Toro in Shizuoka Prefecture have revealed wooden plates, bowls, basins, stands, and ladles. Some of these are lacquered pieces. Forms vary from circular and oval to boat-shaped, all of cryptomeria. Sizes also vary greatly. The wood was gouged and scraped with chisel and ax to create irregular shapes that seem to indicate an as yet immature mastery of the medium. It is likely that these food vessels were meant for home use and were made at home. The people of each household created the forms they desired for containers of their food. Thus these vessels became the symbol of the smallest group unit, the family. They signified all that was needed for the existence of the blood-relation group surrounding the individual.

Clay vessels were used along with the wooden vessels. Their shapes resembled those of wooden vessels, and they fulfilled the same function. Among those peoples whose group life centered upon paddy rice cultivation, vessels were classified according to their respective functions. Such classification came long after the group structure had been conceived and organized. With this classification, fixed life patterns evolved. The type of vessels handed down from one generation to the next remained the same.

Once the actual forms of these vessels became fixed, there arose the new problem of quality. Whether a vessel was of wood or clay became a means of designating and clarifying class distinctions within the group society. By the Heian period the materials used for making food vessels varied greatly. Metal bowls of gold, silver, white bronze, and bronze appeared. Wooden vessels were enhanced with lacquer coatings. As patterns of life solidified, the shapes of the vessels used also became standardized, and the only way to distinguish rank came to be the use of unusual materials or elaborate decoration. This is why metal vessels appear alongside earthenware and ceramic bowls. Bowls could be covered or uncovered, another means of distinction. Plates and other dishes also included silver, copper, wood, earthenware, and other ceramic wares. Some of the more splendid of these are mentioned in such chapters of *The Tale of Utsuho* as "The Open Beach" and "Opening the Storehouse." Although some literary license must be allowed for, these stories are still a valid means of discovering the life patterns of their day. Most of the forms and types of eating vessels of those days are little different from those used today. This is only natural when it is considered that the pattern of a rice-eating culture has continued up to the present.

There were two types of elements in the Japanese life pattern that centered on the act of eating. The first included meals made for *hare,* or extraordinary days, and the second included meals made for *ke,* or ordinary days. *Hare* days were those reserved for ceremonial occasions such as the family and group rituals that might take place in a village community. *Ke* were the days of daily routine existence. On *hare* days menus were varied, and many types of food vessels were used. Only simple, ordinary vessels were used on *ke* days. Japanese groups, beginning with the family unit, usually had communal meals on *hare* days. Eating together at the same time and place helped to establish community solidarity and was a means of reviving and strengthening mutual bonds.

The Communion service of the Christian church has the roots of its meaning in the sharing among believers of the ultimate sacrifice to God. Symbolic partaking of the sacrificial flesh and blood reasserts each believer's self-awakening. In a similar ceremony in Japan believers share in a common meal those edible items that were first offered up to the gods. This is called *naorai.* As in Christianity, it is a way of acknowledging the god common to all and also gives vitality to the bonds uniting the people. The same purpose is achieved in the family meal, but on a smaller scale.

The communal meals that were eaten on *hare* days necessitated a great many vessels and utensils. A set of utensils and tray was developed to meet these needs. During the Bunka and Bunsei reigns of the early nineteenth century the use of these sets spread throughout Japan. Many trays were marked with the family crest, and vessels were coated with red lacquer on the inside and black on the outside. Occasionally a tray might be elaborately decorated with *maki-e,* a kind of gold or silver lacquer decoration. The use of the family crest was just another method of expressing group solidarity.

There is a legend concerning the lending of these sets still told in many parts of Japan. Near a village, the legend says, there would always be a pond or cave in which a dragon lived. When a large number of trays were needed for a ceremonial occasion, the villagers went to the dragon and begged him to lend his sets of trays. Two or three days later the required number of trays and utensils would be found at the edge of the pond or cave. It was important to return the complete sets after their use, but there was one time when a careless man failed to do so. The dragon was angered and forever after refused to lend his dishes. This legend points up the difficulty of acquiring complete sets of ceremonial dishes and trays. When a man was able to obtain a complete set, he would treasure it as a status symbol.

Eating is a form of consumption. As the foodstuffs consumed came to be limited to certain things, the forms of the food vessels also became fixed and standardized. The opposite of consumption is, of course, production. The important thing in production is the great variety of tools and implements necessary.

The Japanese word for tools, *dōgu* (道具), originally referred to the implements used in Buddhist ceremonies and had religious connotations. *Dōgu* were those things needed to follow the Way, or Path (道). Gradually the meaning of *dōgu* was expanded to cover the weapons of the samurai class.

Weapons are extraordinary objects not needed in daily life. They hold the key to life and death. The higher the quality of the weapon, the better a man is able to defend his life and, by the same token, to take life. Weapons are *dōgu* that can shift from life to death in an instant. Their design focuses on effectiveness and utility. The weapon that best fulfills its function is a thing of beauty in and of itself.

Until the advent of modern times, battles were waged on an individual basis, one man against another in a fight for life or death. Here the weapon took on another attribute, for not only might it have beautiful form and be elaborately decorated, but also it could confer upon its bearer magical strength. The unessential decoration was a symbolic prayer evoking magical properties. There are many sword guards decorated with the motif of a dragonfly because another name for the dragonfly was victorious insect. Motifs of tigers, lions, and dragons were all similar attempts to transfer to the warrior the valor and power attributed to these creatures. In China the chrysanthemum and the peony, both symbols of longevity, were common motifs.

The Japanese name for the layered marks left on a sword blade after forging is *masame,* or straight grain, and this is another interesting point. This use of a biological term for a man-made creation shows how the Japanese have always conceived of form with images derived from nature.

Eventually the word *dōgu* was applied to household and personal effects, and tools were developed for their production. Agricultural implements were first referred to as *dōgu* in the *Hyakushō Denki* (Biography of a Farmer) of the Genroku reign (1688–1703). The author of this work is unknown. At the time it was written there were already craftsmen who specialized in the making of farm tools. The author stipulates that "a variety of *dōgu* should be chosen to work with, keeping in mind the benefits to be derived from each. Good *dōgu* are well made, and so the damage they may incur is slight." The text lists in detail the different types of spades, hoes, iron-headed hoes, and sickles that are to be used for different types of soil and crops. The author understood well the need to vary the shape according to its function.

Farming tools held the same importance for the farmer as did the bow and arrow, spear, gun, halberd, and sword for the samurai, or the tools of production for the craftsman. Just as the samurai and the craftsman were always diligent in the care of their *dōgu,* so the farmer should care for his. If the *dōgu* were to become soiled or rusty, they would no longer help to preserve his existence. The author of the above-mentioned farming manual compared such useless tools to the dust-coated abacus and the measure so important to the merchant.

In the Edo period (1615–1867) society was stratified into four levels, with the samurai at the top, followed by the farmer, the craftsman, and the merchant at the bottom. Each class had its own set of *dōgu* symbolizing its rank in society. As symbols it was important that they always be in good condition. The agricultural tools important for cultivation became moral symbols in this feudal society.

The development of the theory that a form is shaped according to the function it is to perform, however, was relatively new. This theory was formulated in the early nineteenth century by Ōkura Nagatsune in his *Study of Agricultural Implements* (1822). Ōkura compiled listings of the measurements and functions of the different types of farm tools he found in Japan and illustrated his text with sketches. He lists twenty-nine different types of hoes alone and also notes that these implements vary in form even when farming communities are only some seventy miles apart. He thereby affirms that form varies according to the conditions of locality and environment.

Such variation was common to all tools of production. Utility or lack of utility of a tool was determined by the natural features of a region and the character of the material used to fashion that particular tool. It was natural, then, that agricultural tools should vary from place to place. The application of decoration to these tools called for the development of new implements to perform new functions.

Specialization and diversification in agricultural tools can best be seen in the tools developed in eras in which farm work was done by manual labor. There is an amazing variation in function and shape to be seen. No matter how great the variation, however, the main premise still held true that the basic form must derive from the function.

FORM AND RELIGION Whether created for the individual or for the group, whether a byproduct of production or of consumption processes, all forms function to support and maintain, on a practical level, life and society. Broadly speaking, forms evolved from nature and were given shape to meet the requirements of human existence. But forms are not born only to meet practical needs. There is another trend in the development of forms in which forms are created out of mental concepts projected by man onto the realities of nature. Up to this point we have been solely concerned with forms of a utilitarian nature. Let us now turn to those forms that symbolize abstract concepts. Conceptual forms are most clearly evident in a religious space and time. The fundamental nature of Japanese religion is to be found in the indigenous Shintoism and in Buddhism; the role of Christianity is minimal.

Shinto originated as a worship of the gods of nature, of such sources of power as the sun and moon, of grand mountains, and of trees of great antiquity. The spirits of people who had achieved great things or met unusual deaths were also common objects of worship. In the worship of all these things people sought relief from disasters and asked for happiness.

Those parts of nature that were considered divine had to be set apart by clearly defined boundaries between holy and profane spaces. A number of such boundary markers evolved in

Shintoism. The torii, the *shimenawa* (a straw-rope ornament), and the *tamagaki* (a sacred fence) are a few of these symbols. The torii, as noted earlier, marks the entrance to a sacred area. The *shimenawa* defines the boundaries of this space, each locality creating its own special design with fresh rice straw.

It was traditionally believed in Japan that the rice straw that remained after harvesting each fall contained the spirit of the new year's grain. The symbolic spiritual value thus implied made the straw a desirable material for the making of *shimenawa*. The idea grew, and abstract objects made of rice straw became common symbols of the new year; they were felicitous symbols, signs of joy and happiness.

Remnants of an archaic pantheism, giant trees or trees with exceptionally gnarled branches were often set aside as sacred objects to be worshiped. These trees were thought to be the resting places for gods descending from heaven. *Shimenawa* were often tied around the trees, or they might be marked with *gohei:* ritual paper ornaments made of white paper, the color of purity. For special ceremonial occasions a *gohei* might be made of the five colors. The hanging paper strips of the *gohei* are in the very simple form of cut and folded interconnecting pieces of paper. It is a mystical form serving no known utilitarian function.

To be saved from disasters and blessed with good fortune are the wishes of all people. Charms and talismans were made to better convey these desires to the gods. A very simple talisman was one inscribed with the name of a god. The very letters of the inscription were believed to be imbued with the power of that god: a belief that stems from Chinese Taoist teachings. In another custom, crude images in human form were cut out of paper, rubbed over the body, and then thrown into a river. Both the talisman and the human image derived their sacred properties from prayer. With prayer the talisman acquired the power of the gods, and the human figure absorbed a man's troubles, washing them away in the river. The significance of both the talisman and the human image came directly from the form.

Food and wine were included in offerings to the gods because it was thought that the gods would enjoy their fragance and taste just as people did. The *sambo,* or footed ceremonial tray, which held the offerings of food and wine, was especially designed with three legs and made of plain unvarnished wood. The vessels and food containers were of undecorated plain white pottery.

The presence of the gods, around which everything centered, was symbolized by objects of abstract design in which the gods' spirits were believed to reside for a time. Under the influence of Chinese Taoism the Japanese gods were at one time represented in human form, but generally they took on abstract shapes. The form, by its very nature, was sufficient to convey the concept of a sacred presence.

This is why there are so many different shapes to be found in Shintoism. It was necessary to make each shape suit the character of the god it represented. In comparison, in Buddhism, which has become a very conventionalized religion, the divine forms are usually of a standard type. The Buddhism introduced through Korea and China had already become a rigid institution by the time it was absorbed by Japan.

The early Buddhist temples, like Christian monasteries, were built as centers of worship and study for Buddhist monks. Later they became earthly representations of Buddha and of heaven. At this point it was necessary to distinguish the sacred area of a temple from the vulgar outside world.

In Japan the use of gold and silver in architecture is nearly all limited to temples, with the exception of two or three castles. Gold and silver traditionally represent extraordinary spiritual power. Today the altars found in many private homes for the worship of Buddha and ancestral spirits often use gold and silver to mark them as holy spaces within the home.

The Buddhist temple is full of symbolic forms. The decorative altar equipment is one example. Of very little practical use and often excessively decorated, these forms contrive to give the temple an otherworldly atmosphere. The first Chinese Buddhist temples were built along the lines of a palace. A palace or castle was a space in which a man of power would reside. It was a symbol that set the man apart from the common masses, and everything in its construction spoke of this. The complex architectural form of the palace was adopted by the Buddhist temple to serve much the same purpose.

The world of religion includes many rituals and ceremonies. Each ritual or ceremony is an enactment of a drama: the drama of a god descending to earth to bring happiness to his people, or the drama of salvation through the grace of Buddha. The instruments and objects used in these rituals and ceremonies serve as links to the spiritual world, bonds between the sacred and the profane.

With the completion of a ritual and the departure of the gods, dramas and plays meant for the entertainment of men could begin. Masks were used in the earliest known of these dramas. Gigaku and Noh are two types of Japanese masked drama. The masks used in these plays represented a dream world, an existence not of this world. Since very human men were enacting the roles, however, the masks could never depart too drastically from human representation. Exaggerated emphasis was placed on various portions of the human face, and in this manner the masks also served to link man to unearthly realms. When a man wore one of these masks, he was no longer just a human being but was privileged to enter the world of the gods. The mask served to sanctify the man.

Many masked dramas centered around ghosts. Noh is an example of this, for in Noh dramas spirits from the underworld are common protagonists. The Noh mask has the unique characteristic of expressing a multitude of emotions simply by being shifted in the light, even though it normally appears to be expressionless. The masked play is a drama of illusions. It should be strongly emphasized that this is the exact opposite of modern realistic drama.

Dramas employing dolls instead of human actors had the same sacred quality. Dolls were thought to have a mysterious life of their own. Ghosts might also reside in dolls. Dolls of this nature can still be seen today on special vehicles, or floats, used in certain ceremonies.

The original plays that employed dolls, like the ancient masked dramas, were means of divination and soothsaying. This tradition is still preserved in the Oshirasama found in northern parts of Japan. The crude human figure known as Oshirasama was made to dance by shrine maidens. As the dancing progressed, spirits were believed to possess the maidens and speak through them. The puppeteers who wandered around Japan in the middle ages probably performed a similar function. These dolls and puppets were very crude, only hinting at their human models. A superficial resemblance was all that was necessary to communicate with the gods. Such figures do not become truly realistic until they are shifted from the world of the gods to the world of man. The early dolls were ritualistic forms wrapped in mystery.

They were not used to act out literary themes until well after the maturity of an urban culture in the Edo period. The city was a man-made world in which the manufacture of objects predominated. With such an environment, dramas gradually came to portray the secular human world. People were drawn to this new form of the drama, and the theater therefore strove to satisfy aesthetic values.

Another name for *homo sapiens* (man as a creature of intellect) is *homo ludens* (man as a creature of leisure). This aspect of man's nature is most strongly manifested in the city, where many forms are created solely for play. The pattern of development is easily seen in the art of flower arrangement and the tea ceremony.

Ikebana, or flower arranging, evolved from the offering of flowers made before the Buddhist altar and in other sacred areas. Later, flowers were used for decoration on days of *hare*—for example, as when a samurai went off to war or a bride to be wed. Ikebana also marked specially designated days of the year, such as the seventh day of the seventh month. Eventually this particular amusement was no longer confined to *hare* days but became an integral part of ordinary *ke* days as well.

Containers and vases are a crucial part of ikebana. It was important for the container to match the theme of any given flower arrangement, such as "heaven," "earth," "man," "yin-yang," or "the five elements." If a flower arrangement was meant only to please a man's sense of beauty, then the container had to do likewise. The earlier flower containers were of bronze and porcelain imported from China, but eventually bamboo sections, baskets, and ceramic wares were also adopted for use. The focal point of ikebana has always been the arrangement of flowers. Ranking second in importance, the flower container underwent very little innovation in design.

In comparison the tea ceremony seemed to be constantly discovering and creating a multitude of new forms. Originally, the ceremony consisted of a kind of game or drama focused on the brewing and drinking of tea. The use of fire and water tied the ceremony to a more mundane reality of man's existence, his eating life. The tea ceremony, however, removed the act of drinking tea from the range of ordinary eating habits, placed it within the narrow confines of the tearoom, and built a great drama around it. In its earlier stages the ceremony was conducted in the ample space of the main room of a house, but Sen Rikyū, in a revolutionary move, changed this to an independent separate enclosure apart from the house.

The utensils of the tea ceremony had to be adapted to the extremely small space, and this innovation gave rise to the development of many new forms. The designing and creation of objects solely for use in the tea ceremony has become one of the more peculiar characteristics of Japanese culture. The Raku tea bowl, for example, was meant exclusively for the drinking of tea in the tea ceremony. The tea caddy, or *natsume,* a lacquered container for the tea powder, and a special type of shelf were all created for the tea ceremony. Other forms developed for the tea ceremony include such bamboo items as flower containers, tea scoops, and ladles. Variations in shape were devised to suit personal taste.

All these forms fitted into a single process, a single drama, independent of secular boundaries and moving within a very limited space. The utensils were chosen for the very reason that they would not be suitable for everyday use. At the same time each piece, each utensil, was a testimony to the personal tastes of a particular tea master, or of the man who made the piece. Thus the forms of the tea ceremony fulfilled two functions: they helped to create a certain dramatic atmosphere and they recalled famous figures in history.

There are no other instances in Japan in which so many forms have been created solely for a leisure pastime. Generally an object made for leisure or play had no practical function and therefore naturally created a time-space cut off from the time-space of daily existence. It was designed to please man within the boundaries of a world of play. As *homo ludens,* man had thus successfully transferred the physical activity of play into a highly abstract conceptual form. He made the very activity itself an object to be designed and created, and he manufactured unique tools to supplement the activity. This is the greatest creativity to have been evolved by the Japanese.

Forms are inherent in nature and in culture. They are an expression of natural forces and cultural achievements. In this sense forms are universal.

There is, however, a very fundamental difference that reveals itself as we compare the geometric conceptions of Vitruvius with the concepts of the Chinese *Chou Li.* The comparison forces us to face another aspect of forms. While many forms seem to be common to all cultures, the majority represent a particular ethnic group's history and mark the gradual development of unique cultures. Therefore it can be stated that forms are strongly regulated by historical considerations. The Greek and Roman temples were contructed of stone in accordance with strict geometric rules. The temples and shrines of Japan, such as the Grand Shrine of Ise, are made of wood, and though they are built with a compact symmetry, their dimensions convey a broad, tranquil atmosphere. Though both styles are meant to create a sacred area, they differ because of the different history of each society. These unique forms testify to the vitality of the respective cultures. Thus do forms symbolize the unique character of every culture.

Both the temples of stone and those of wood are limited by the nature of their material. Europe developed an architecture of stone, and in western Asia brick was used. In Japan the medium was wood. One type of architecture was based on walls, the other on pillars. Japan's warm and humid climate ensured the development of a culture whose forms would be made of natural vegetable materials. The natural environment determined the type of forms that would be created.

Man has two histories: as an individual and as a group. The forms developed by both encompass the history of the whole human race. Man, as the creator of history and the builder of societies, perpetually gives birth to new forms.

Employing his hands and tools, man has endeavored to dig out the forms he sees sleeping within natural materials. The form exists first as a concept that later takes on concrete shape as the process of creation occurs. Forms are a declaration of man.

Forms exist for every one of the different facets of man's character: his character as *homo faber* and as *homo ludens*. It is the principal *modus vivendi* of man to recognize and deal with the forms latent within his being.

WOOD

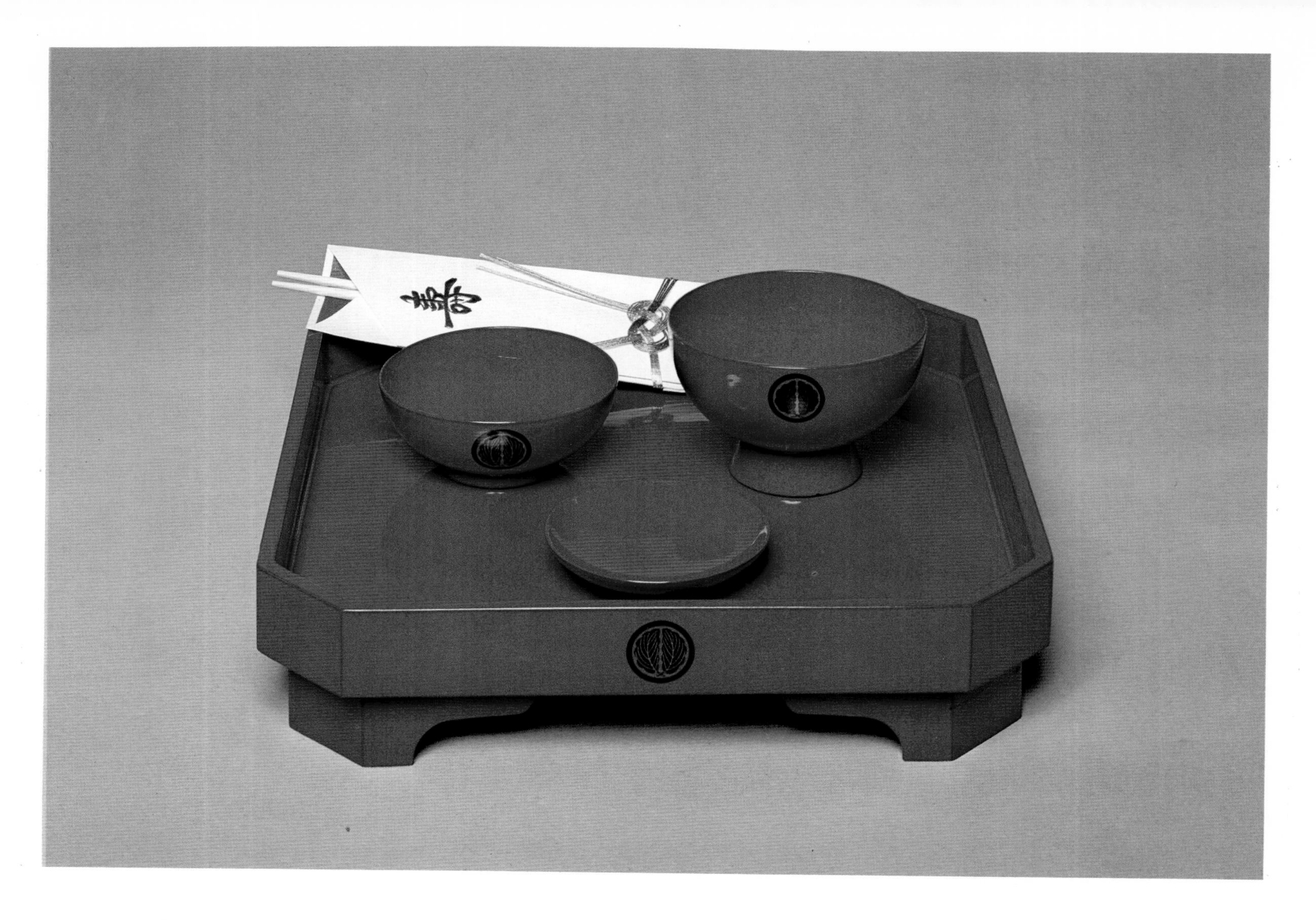

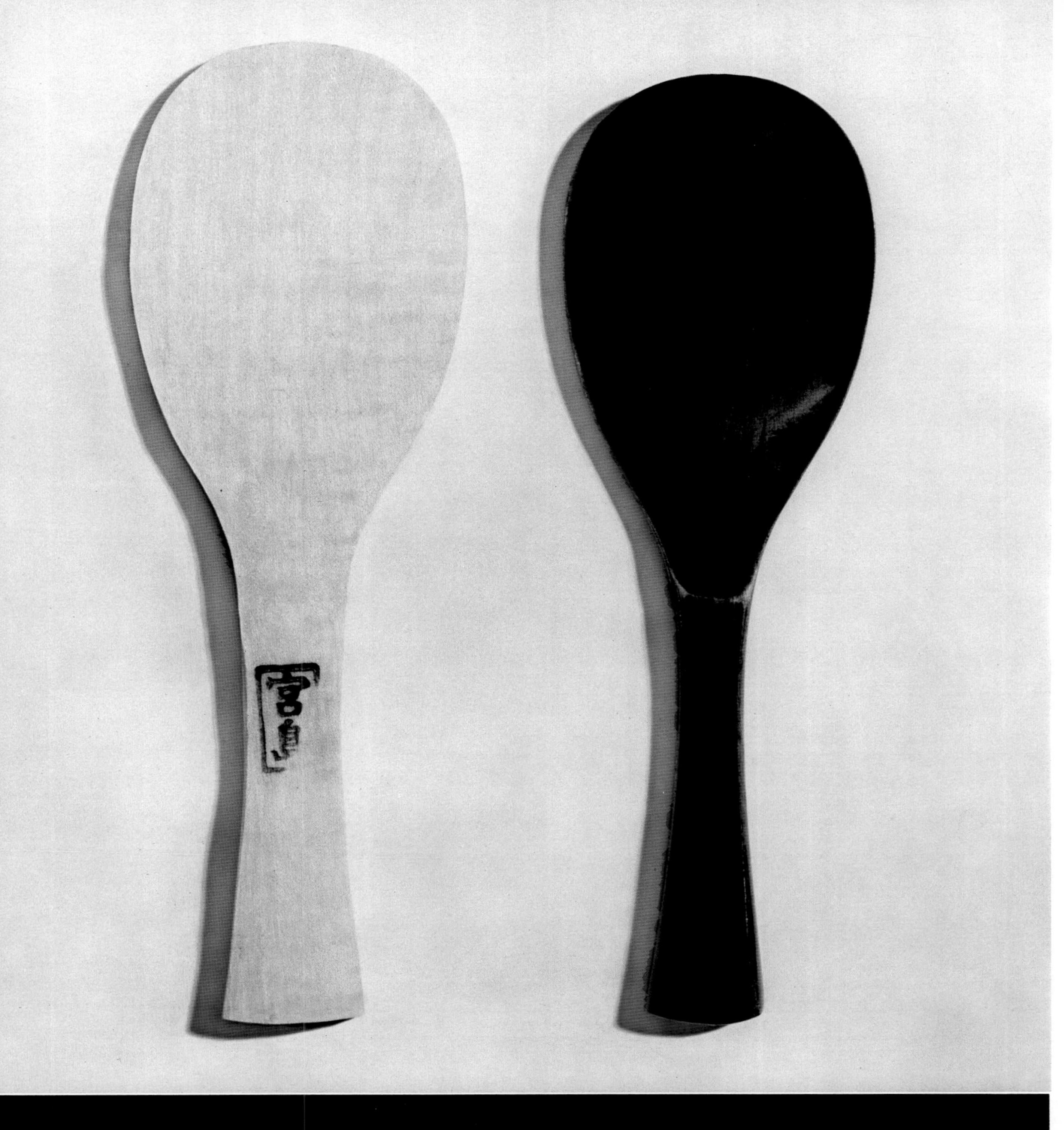

液用
一升

穀用
一升
約一八リットル

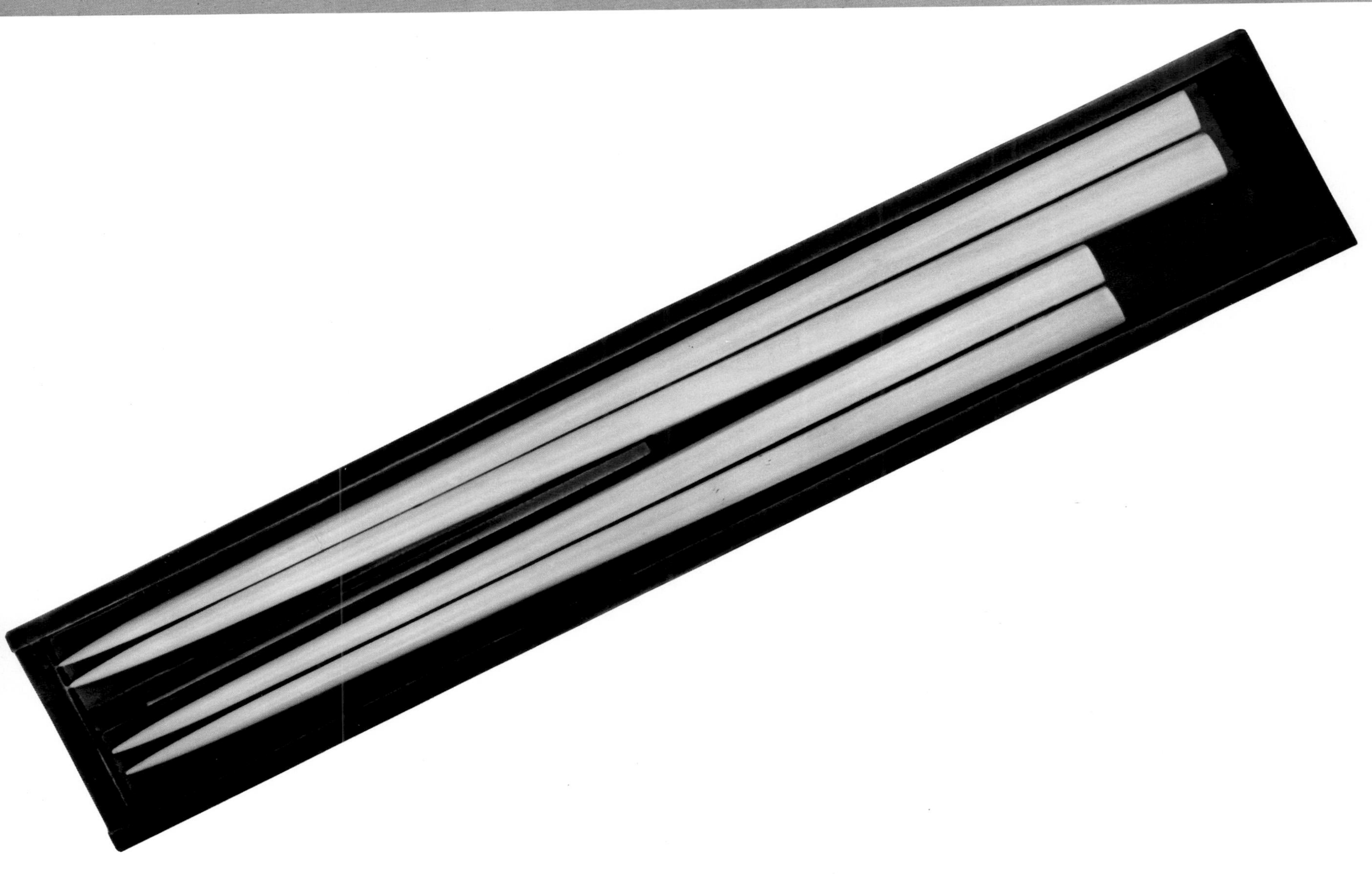

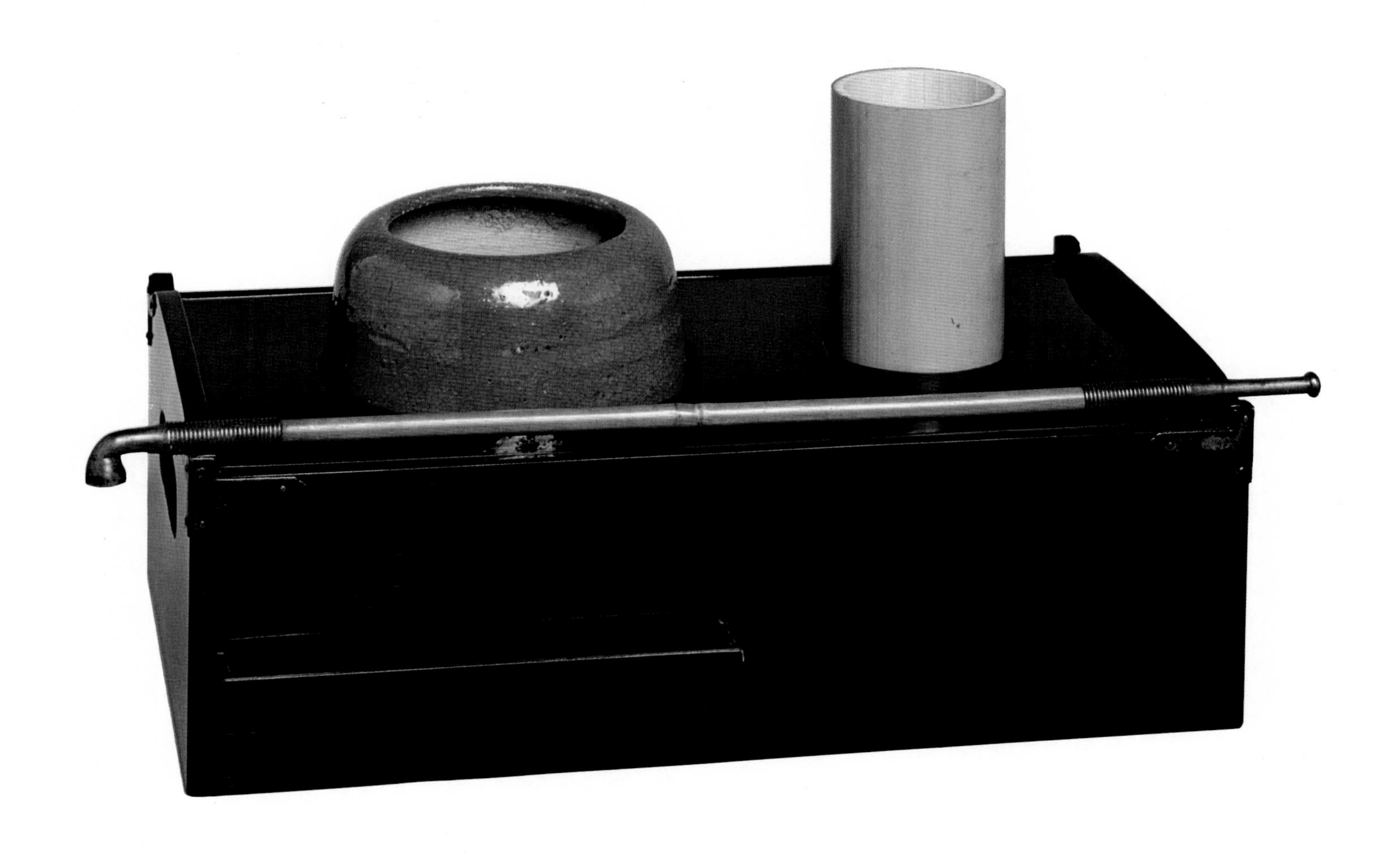

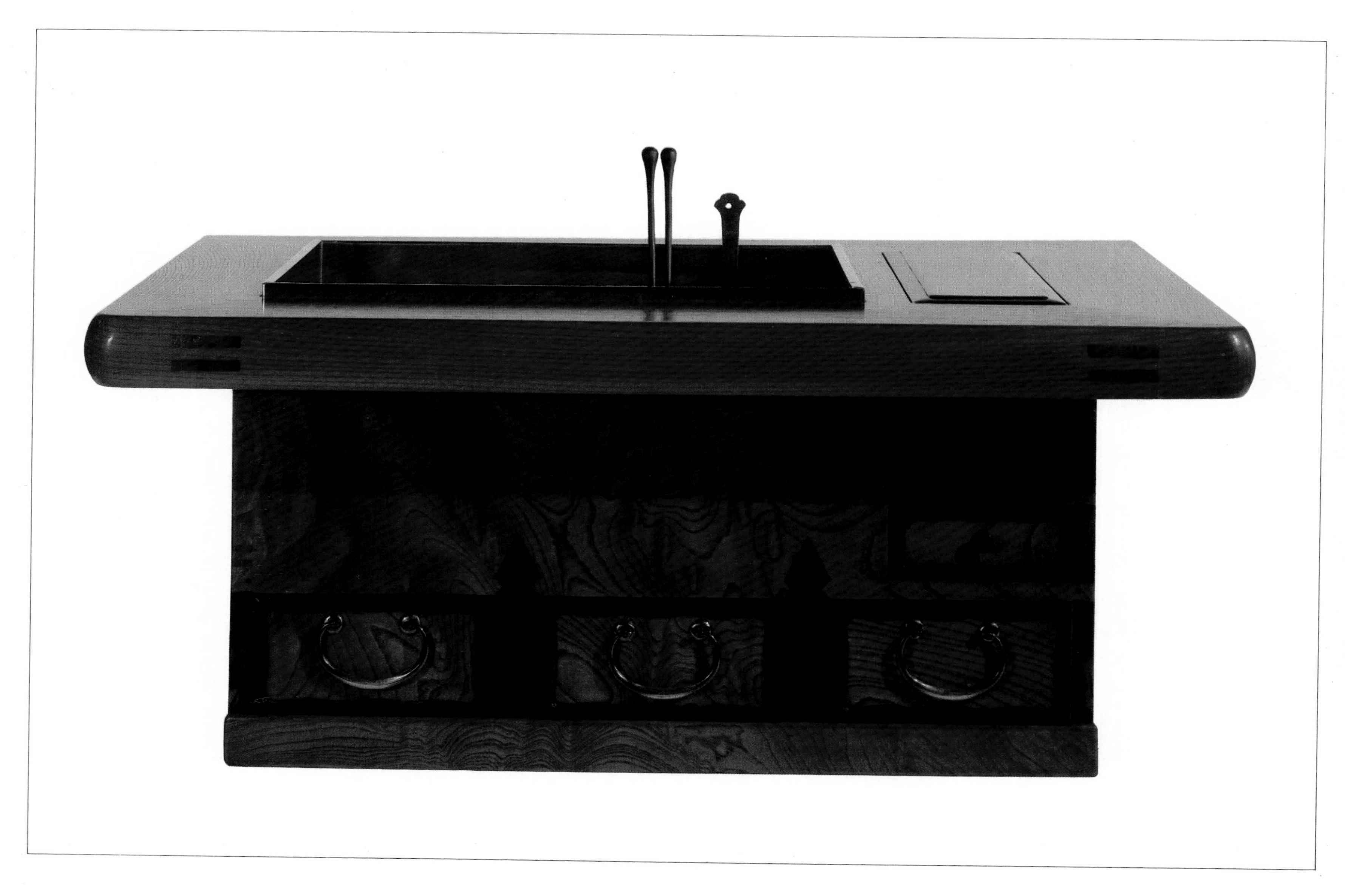

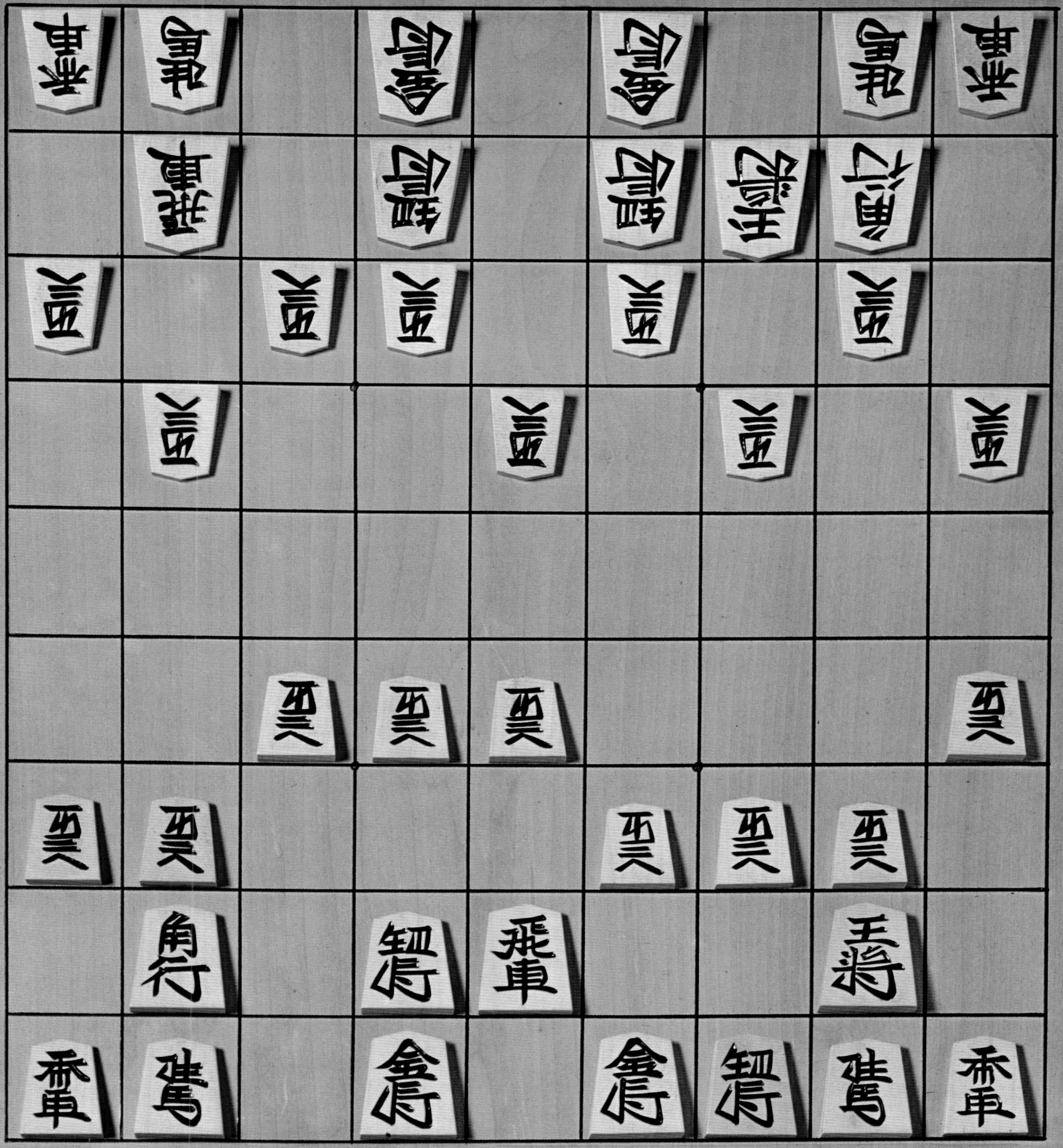

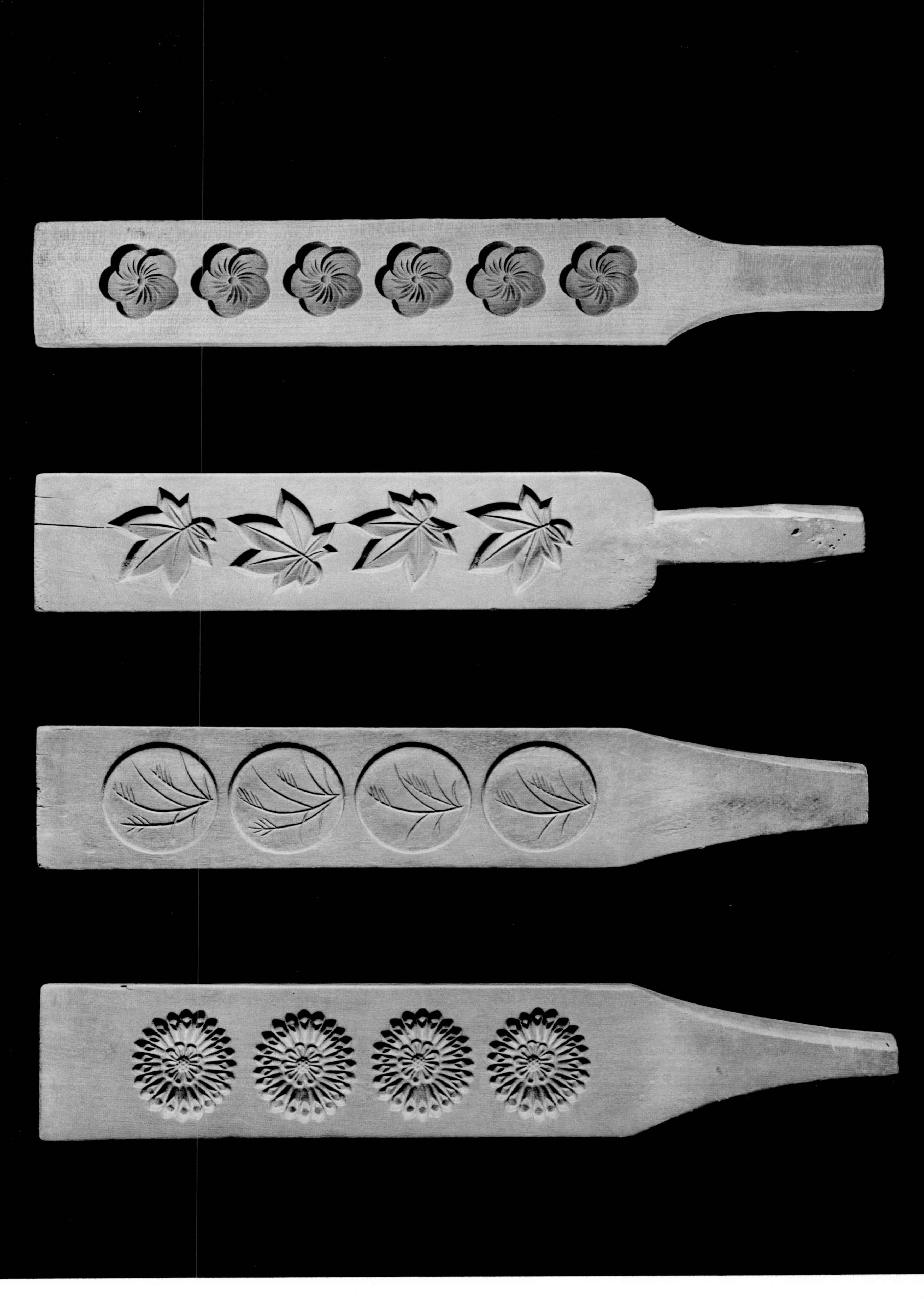

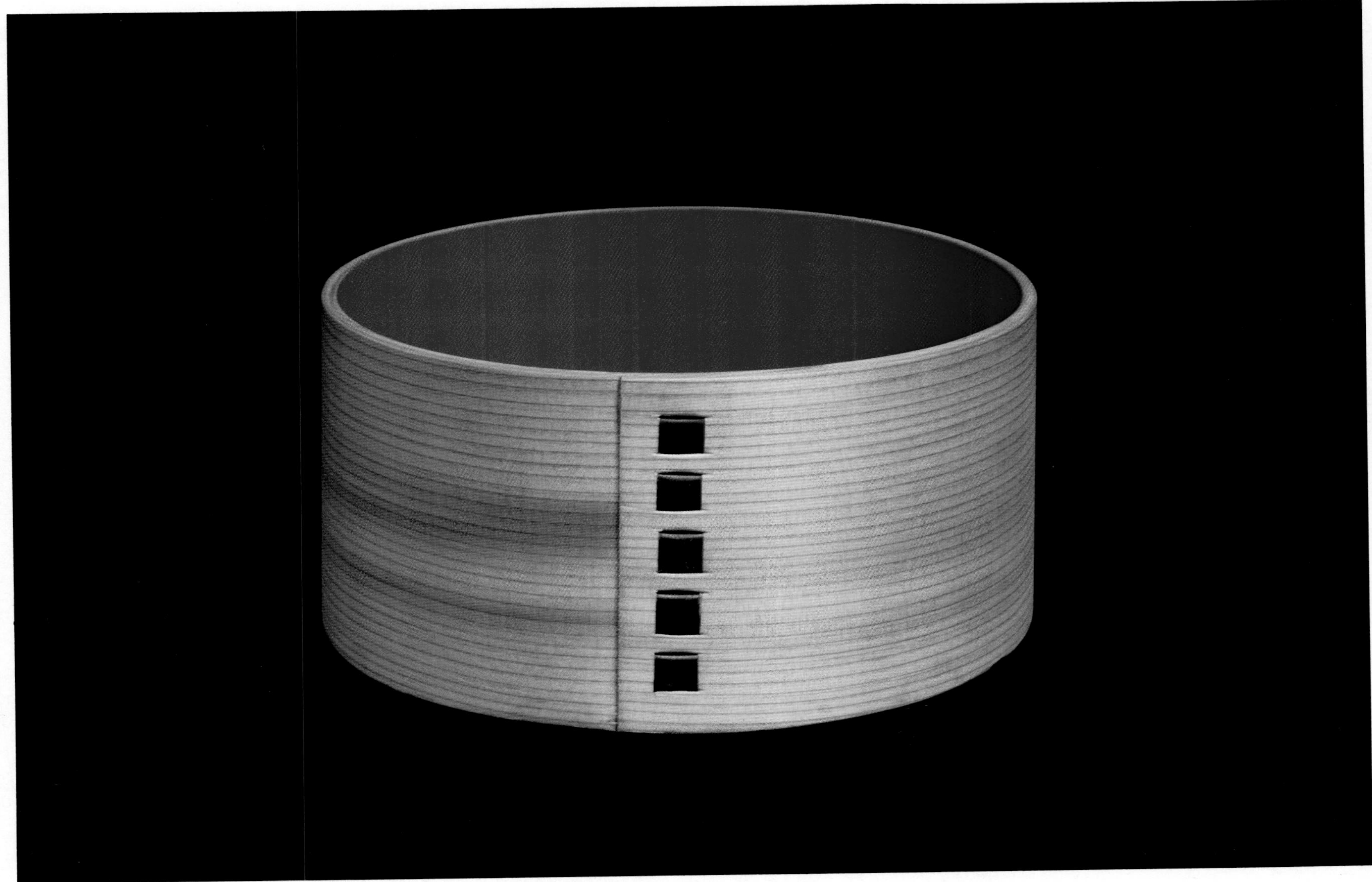

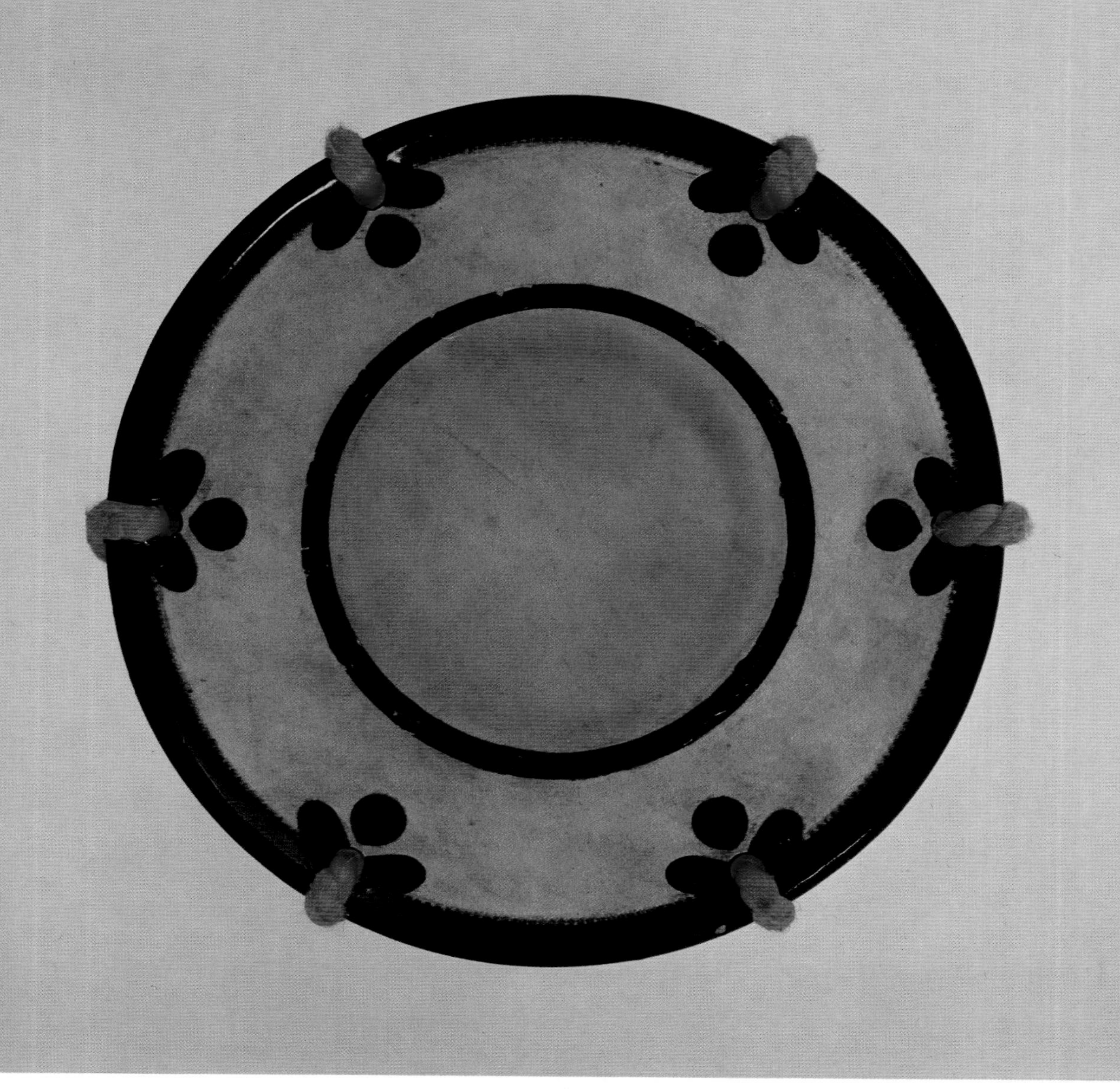

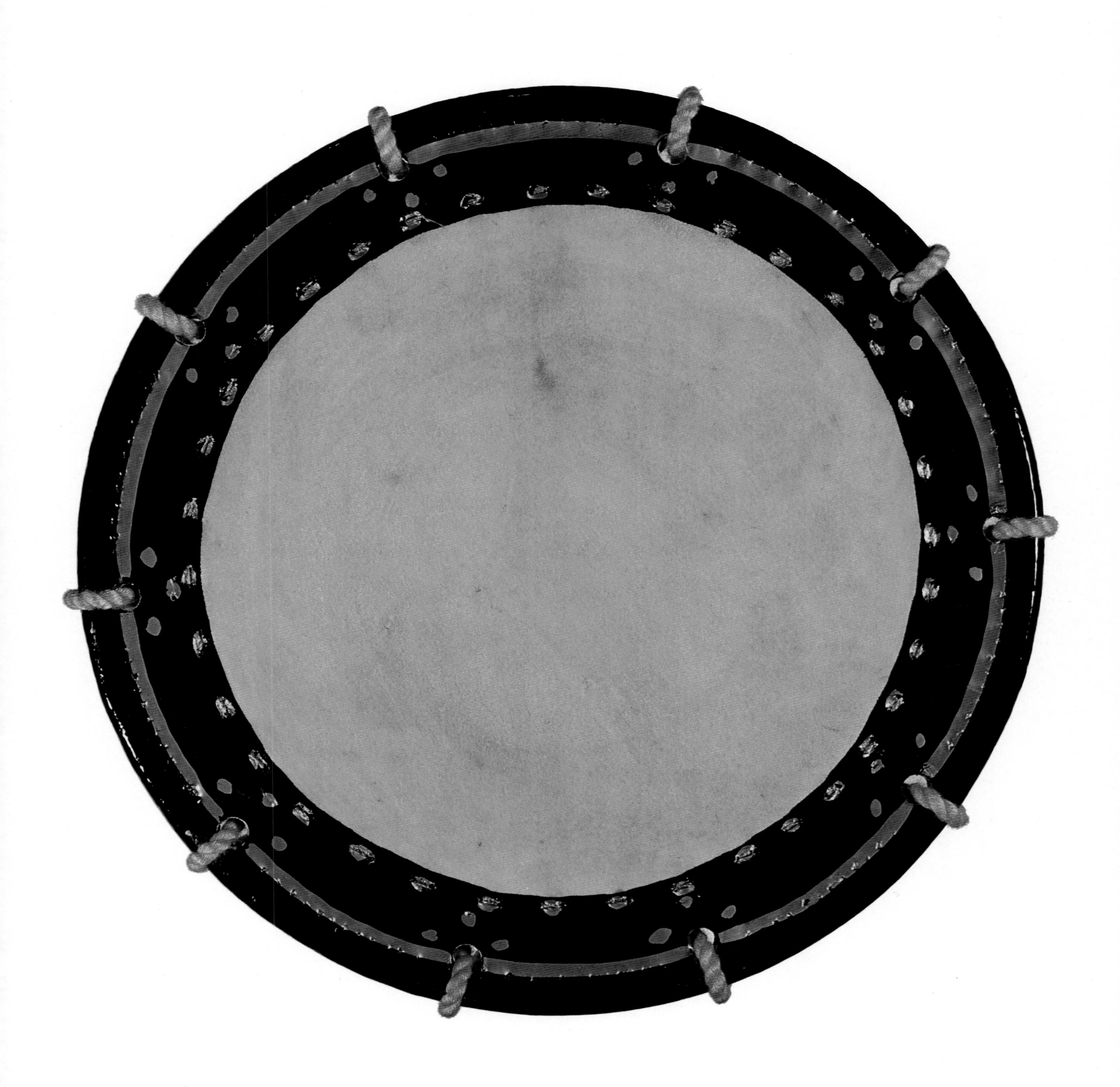

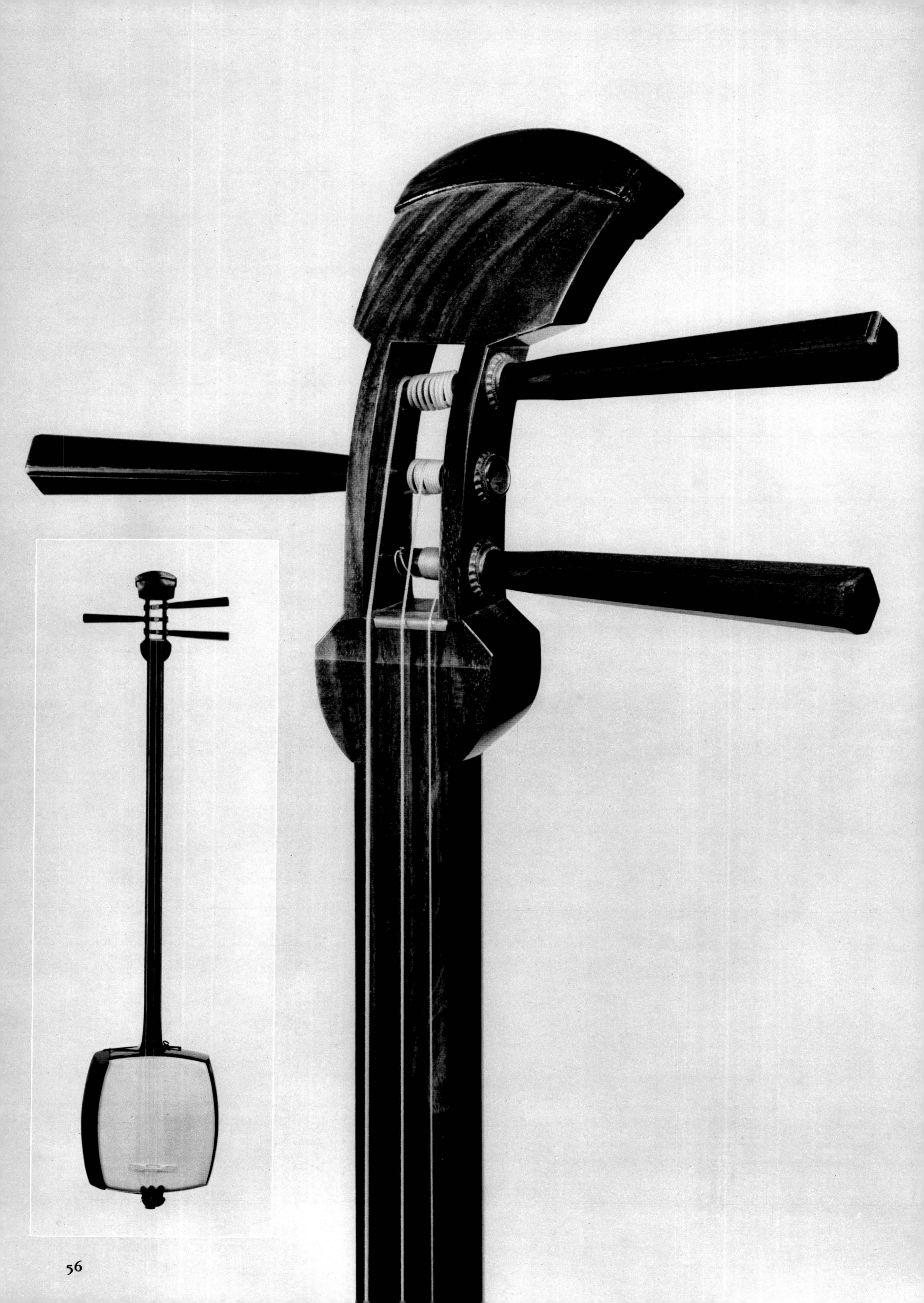

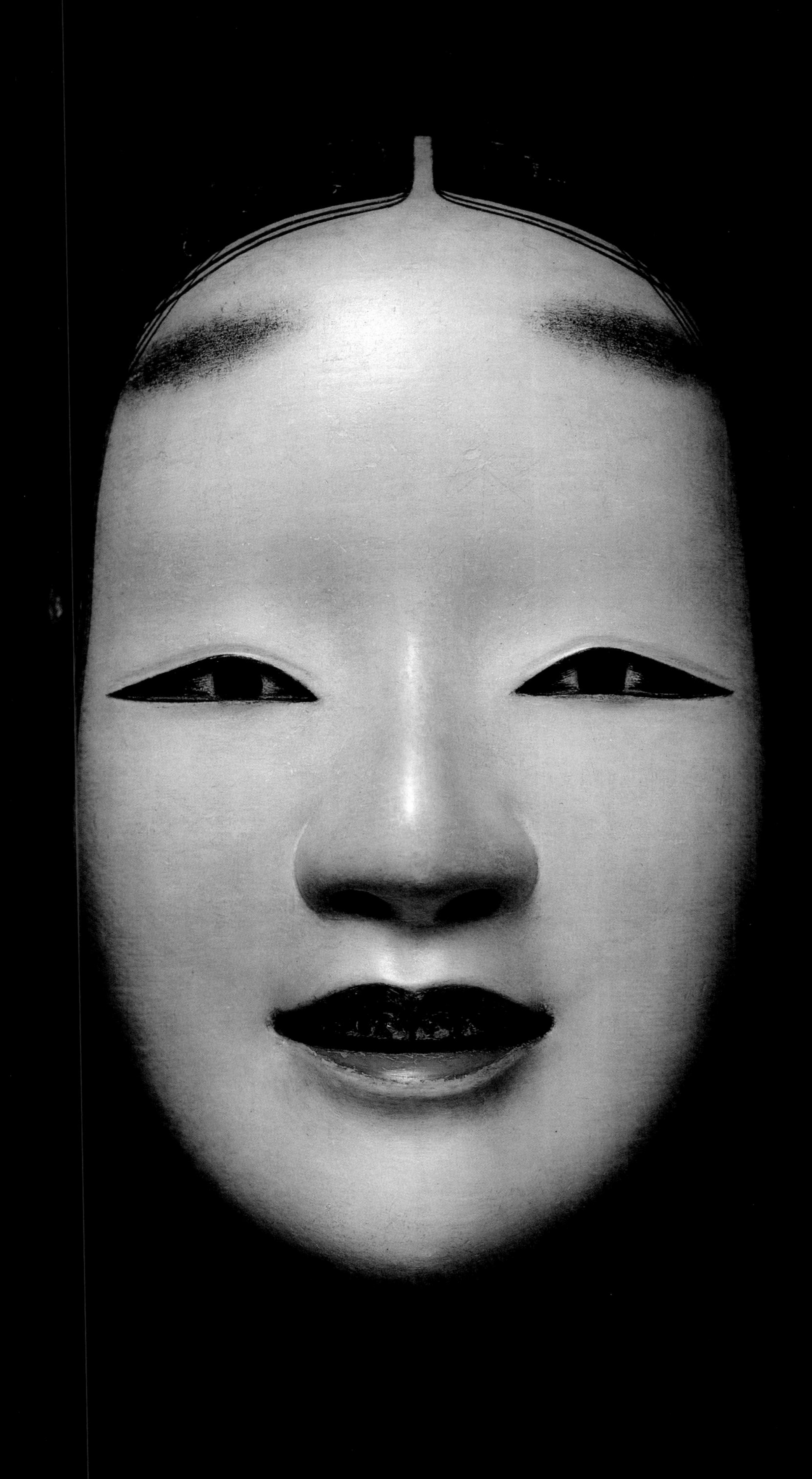

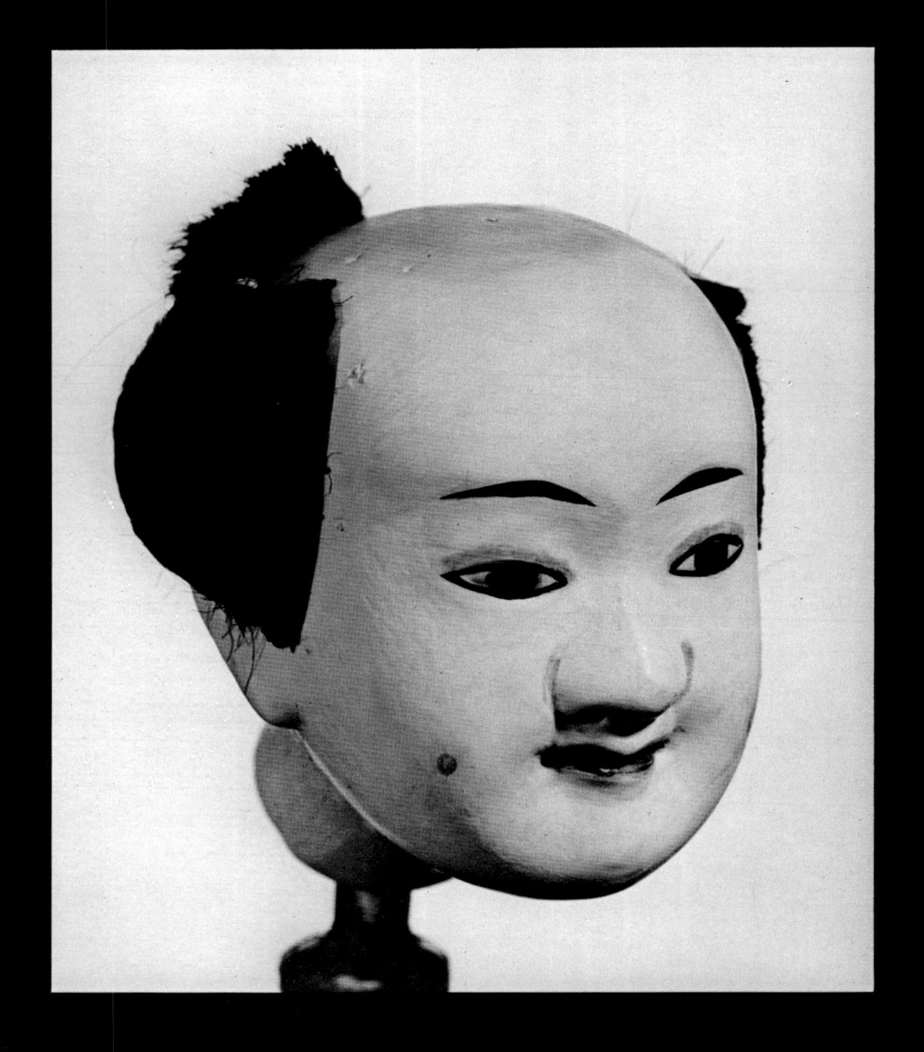

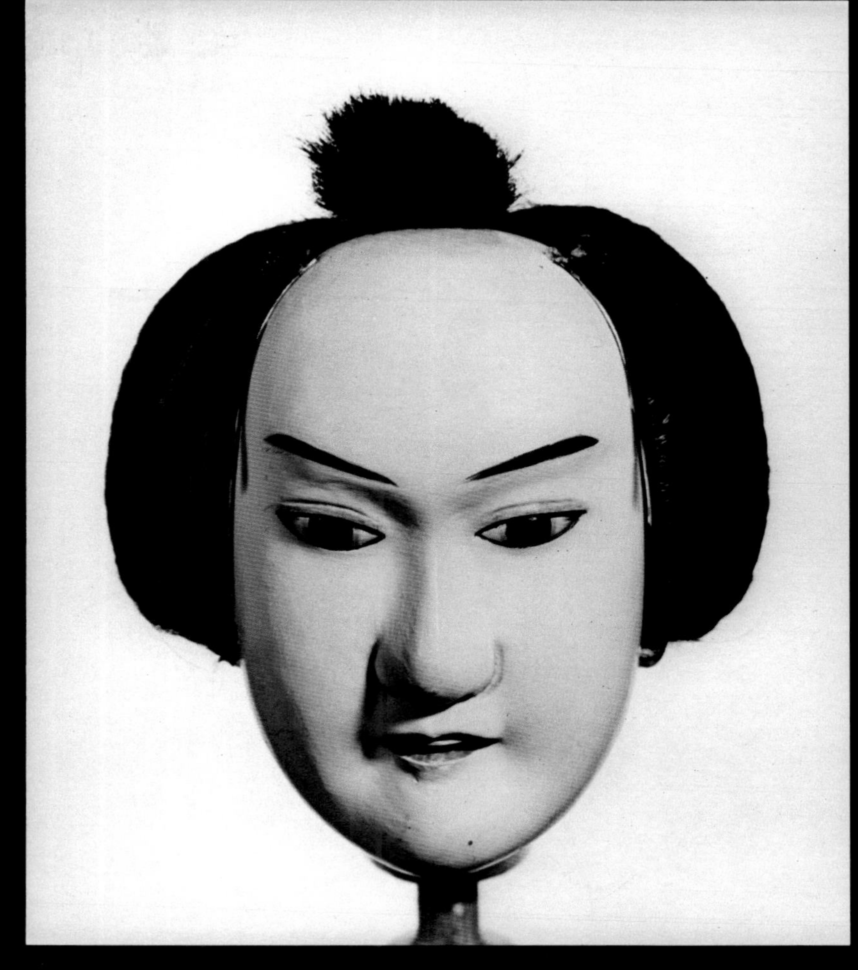

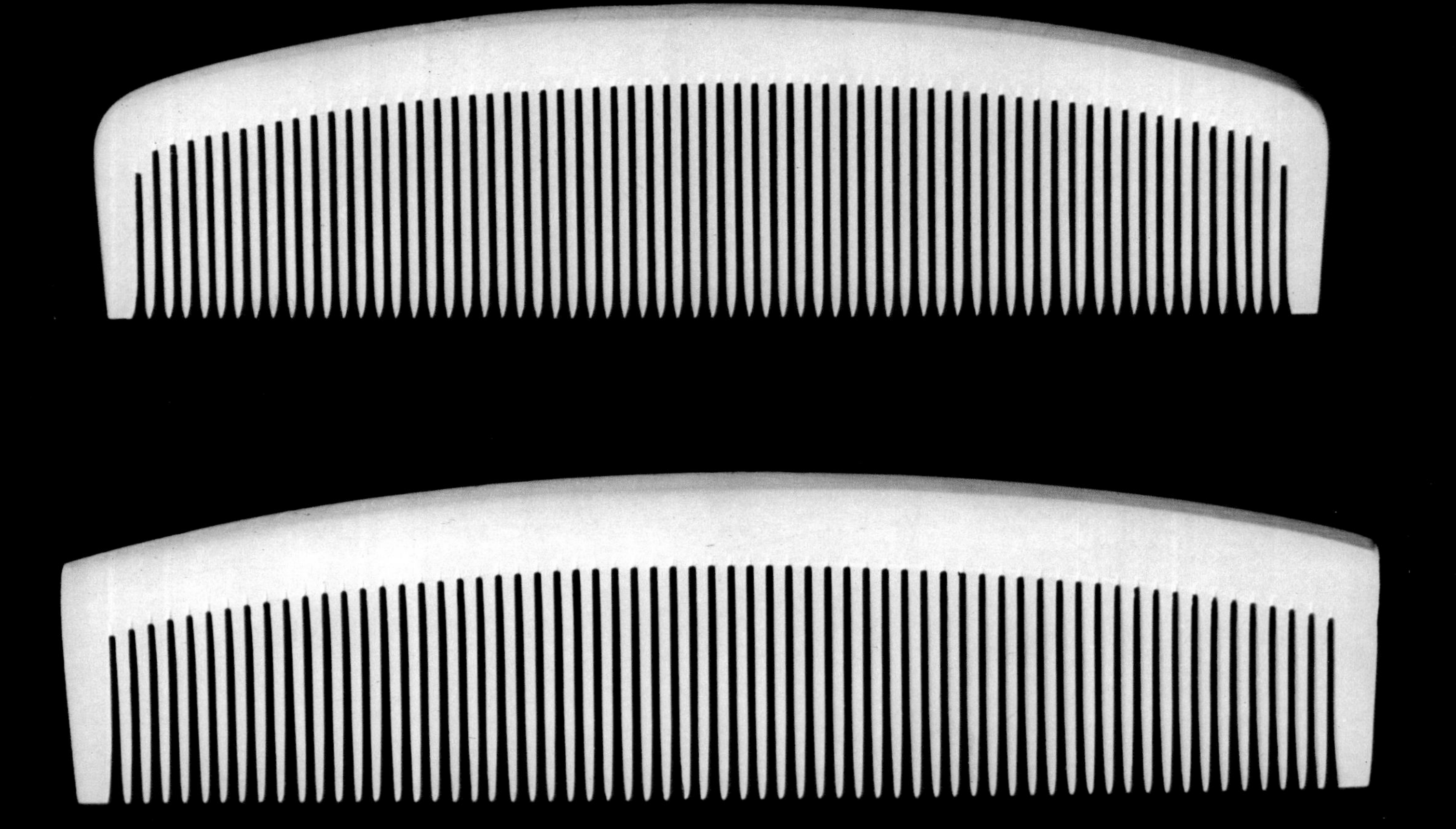

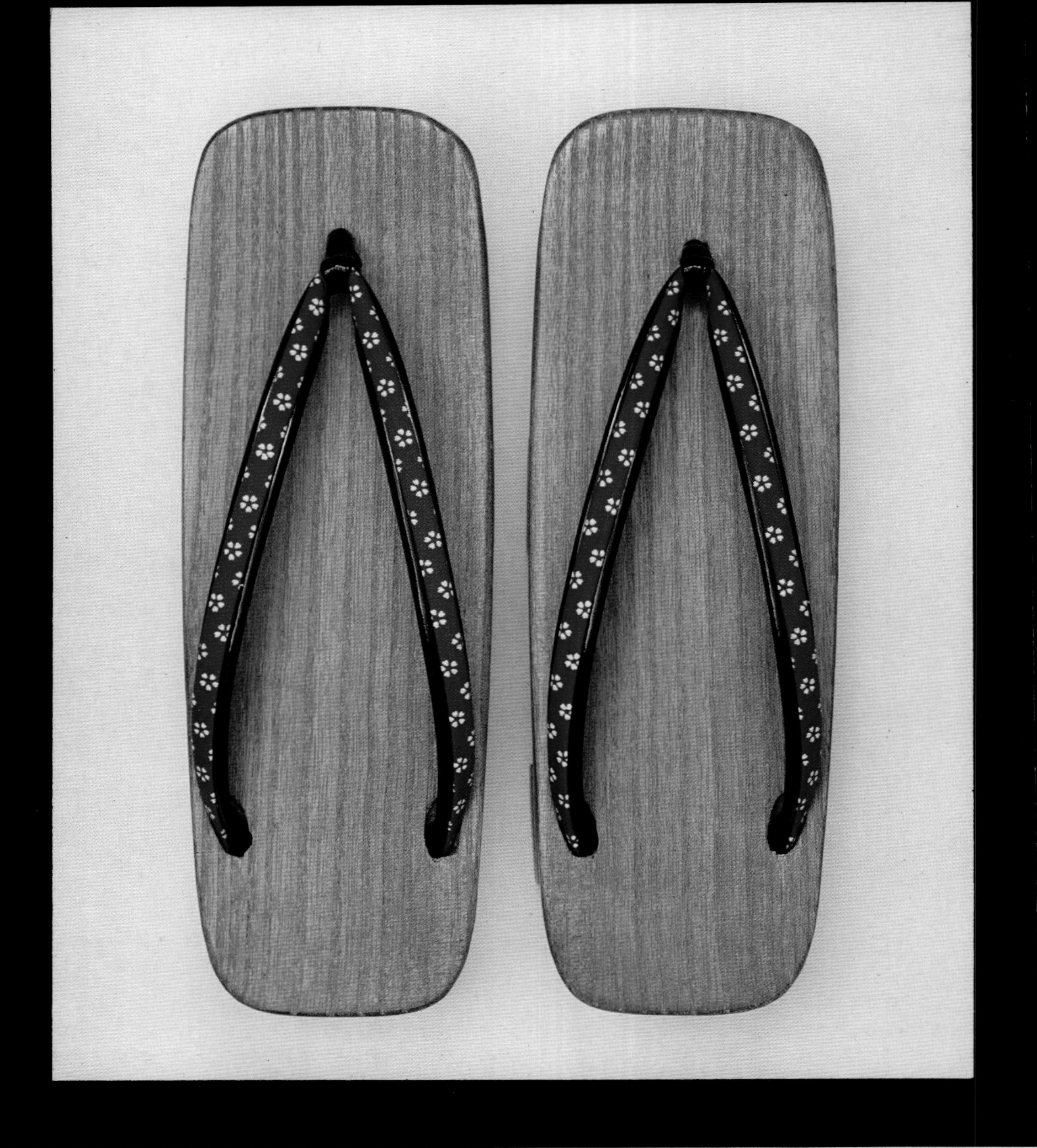

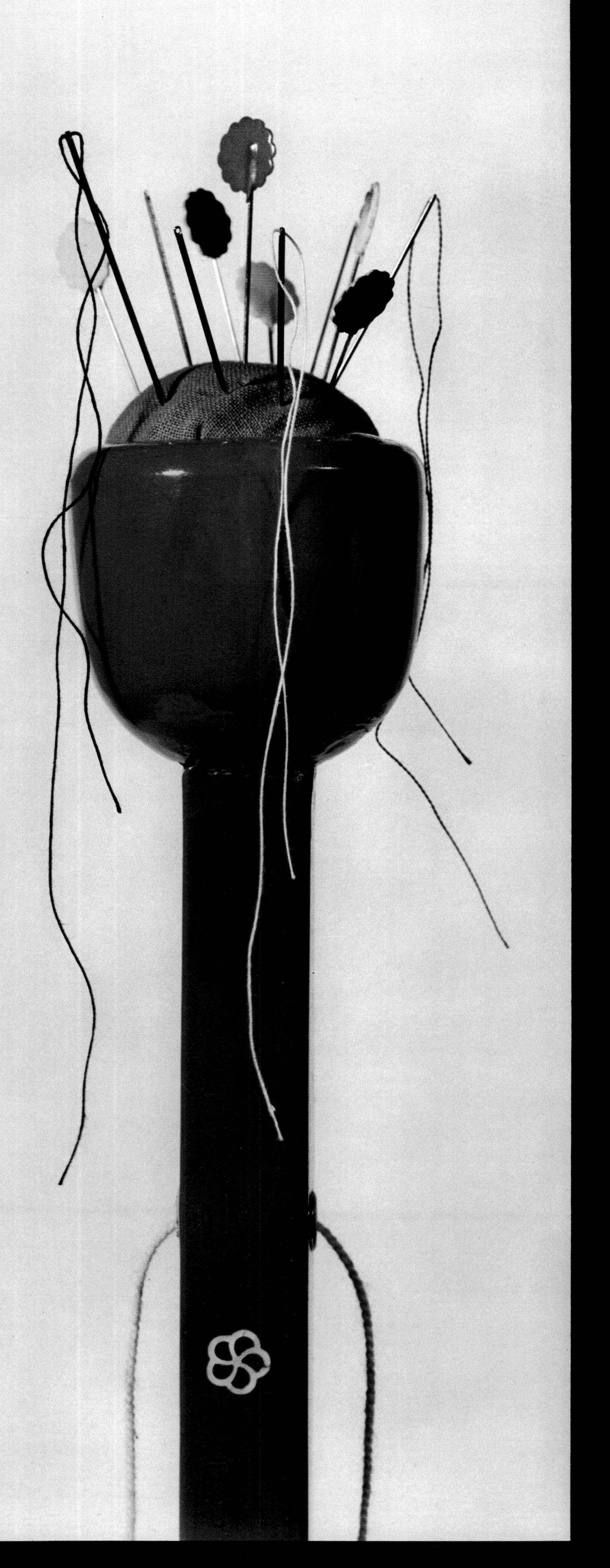

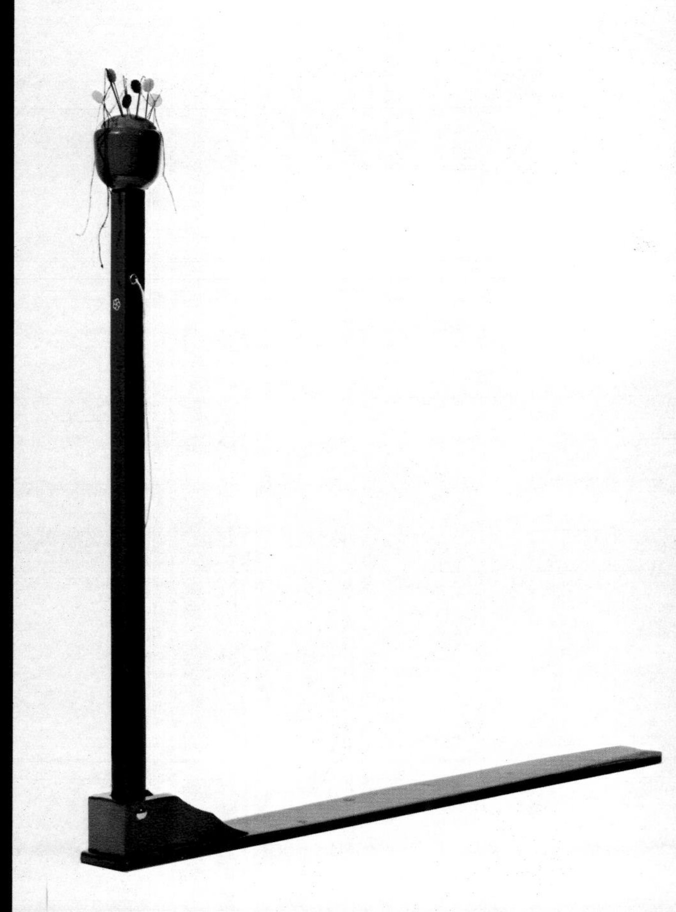

奉 末広大神 納
昭和三十七年
奉 岩本大神 納
奉 末広大神 納
昭和三十七年一月吉日建之
今峰忠男
綿善商店

納 神大森王
奉 阪急野田駅前
昭和三十七年
中島光夫

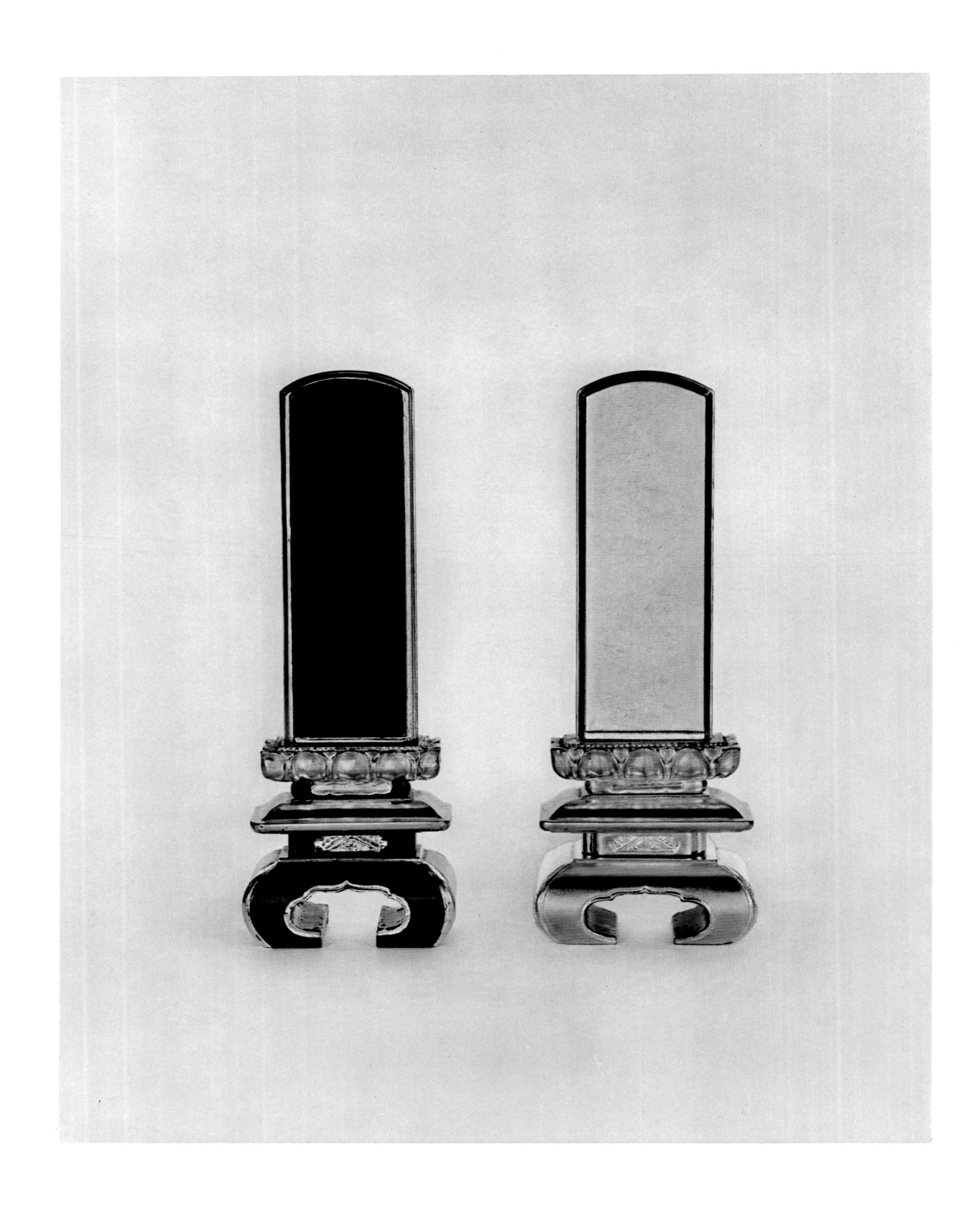

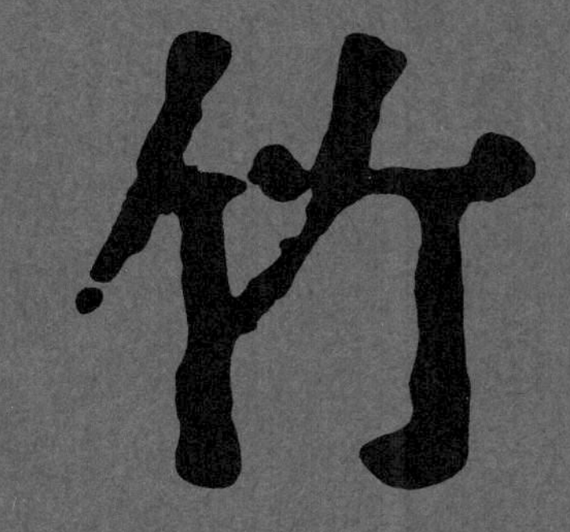

BAMBOO

81

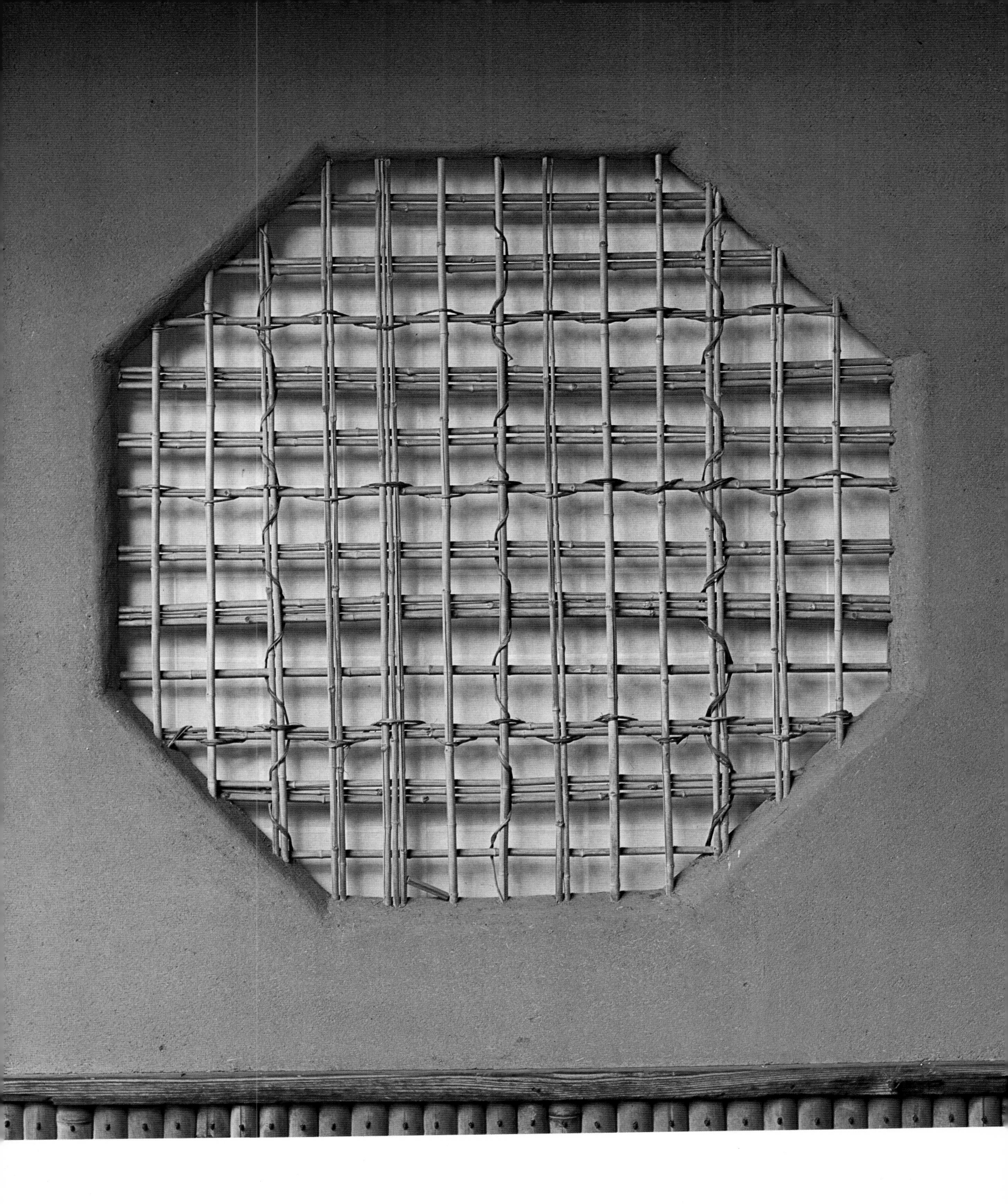

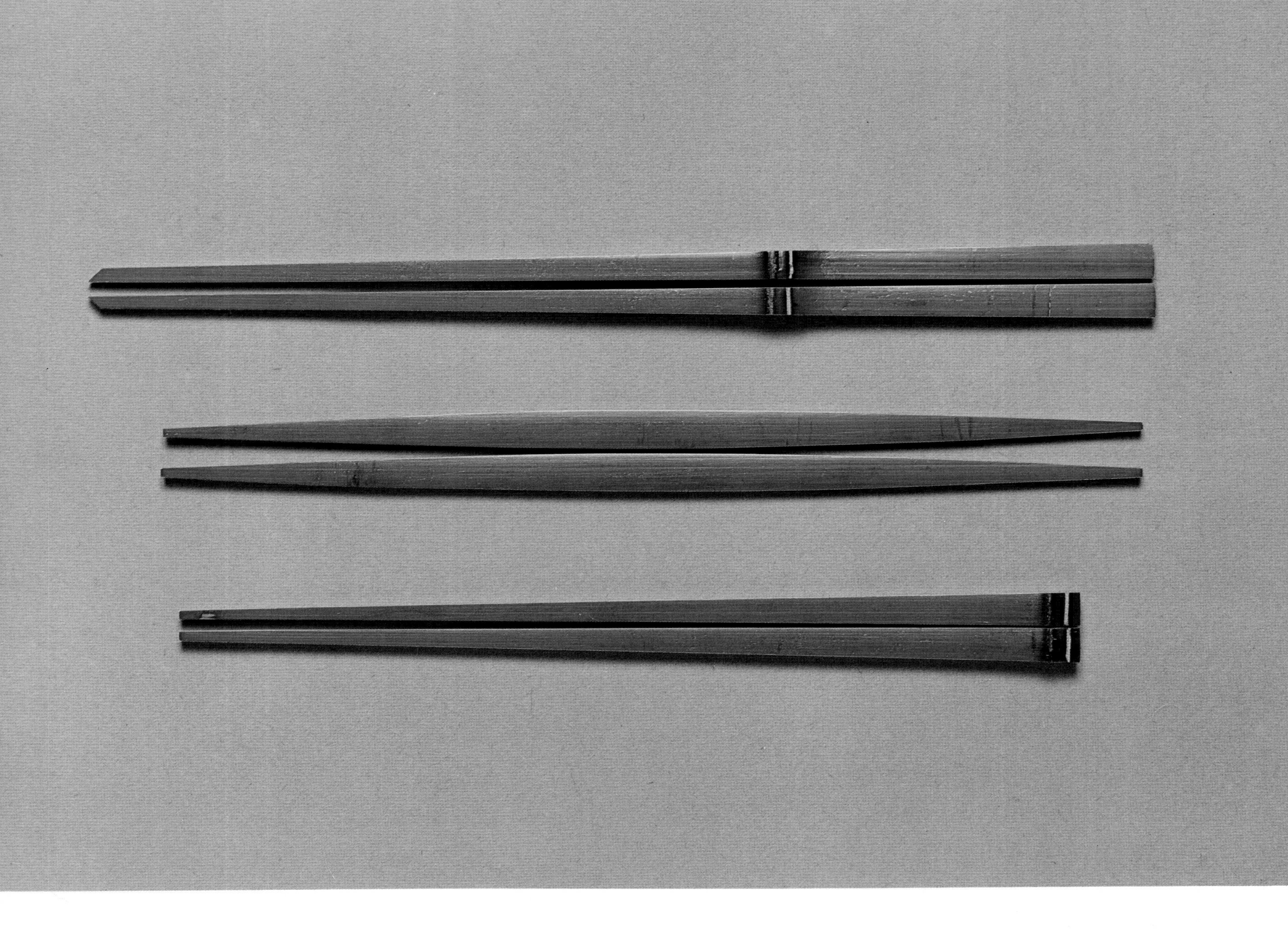

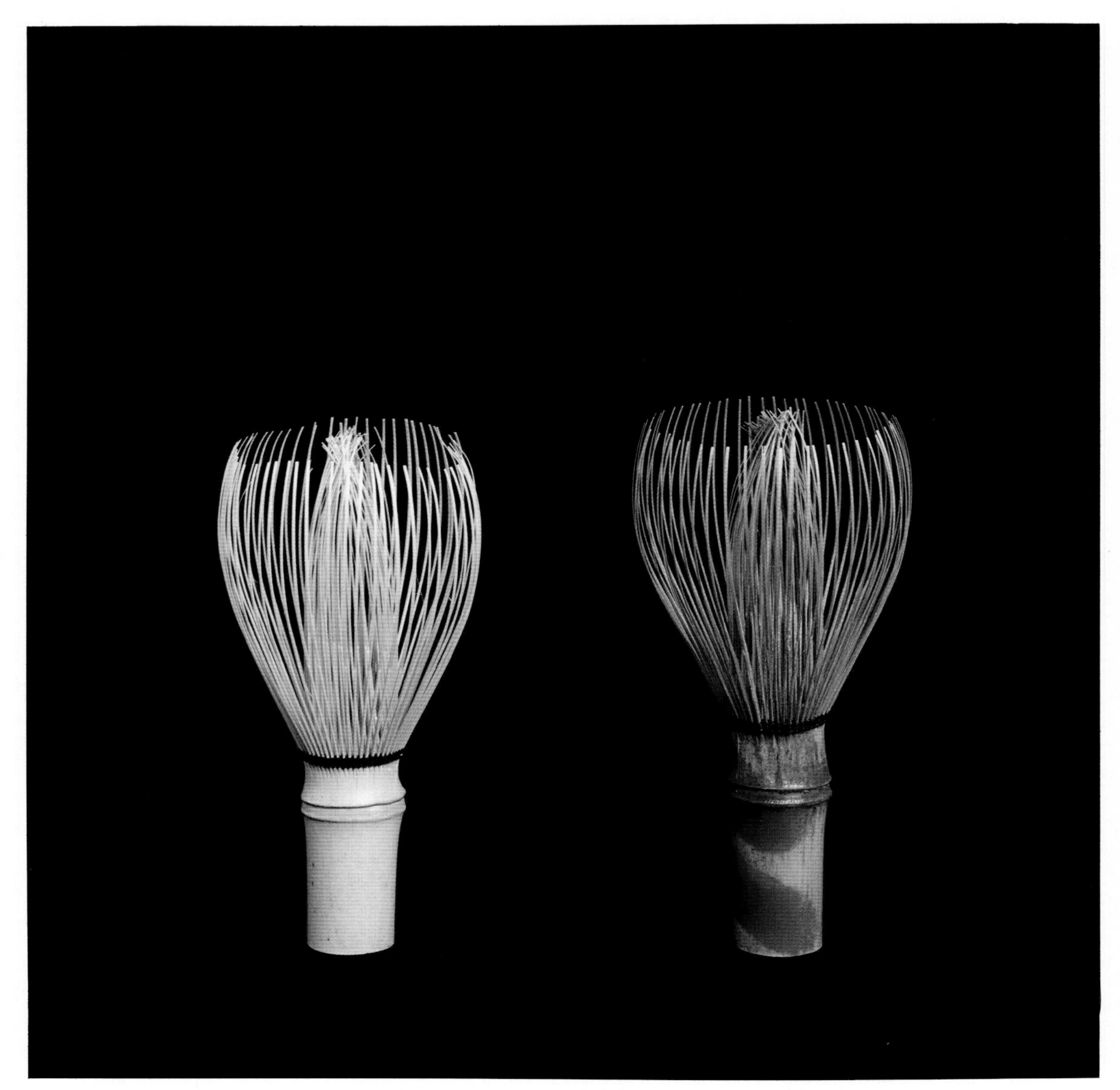

帯屋

献燈

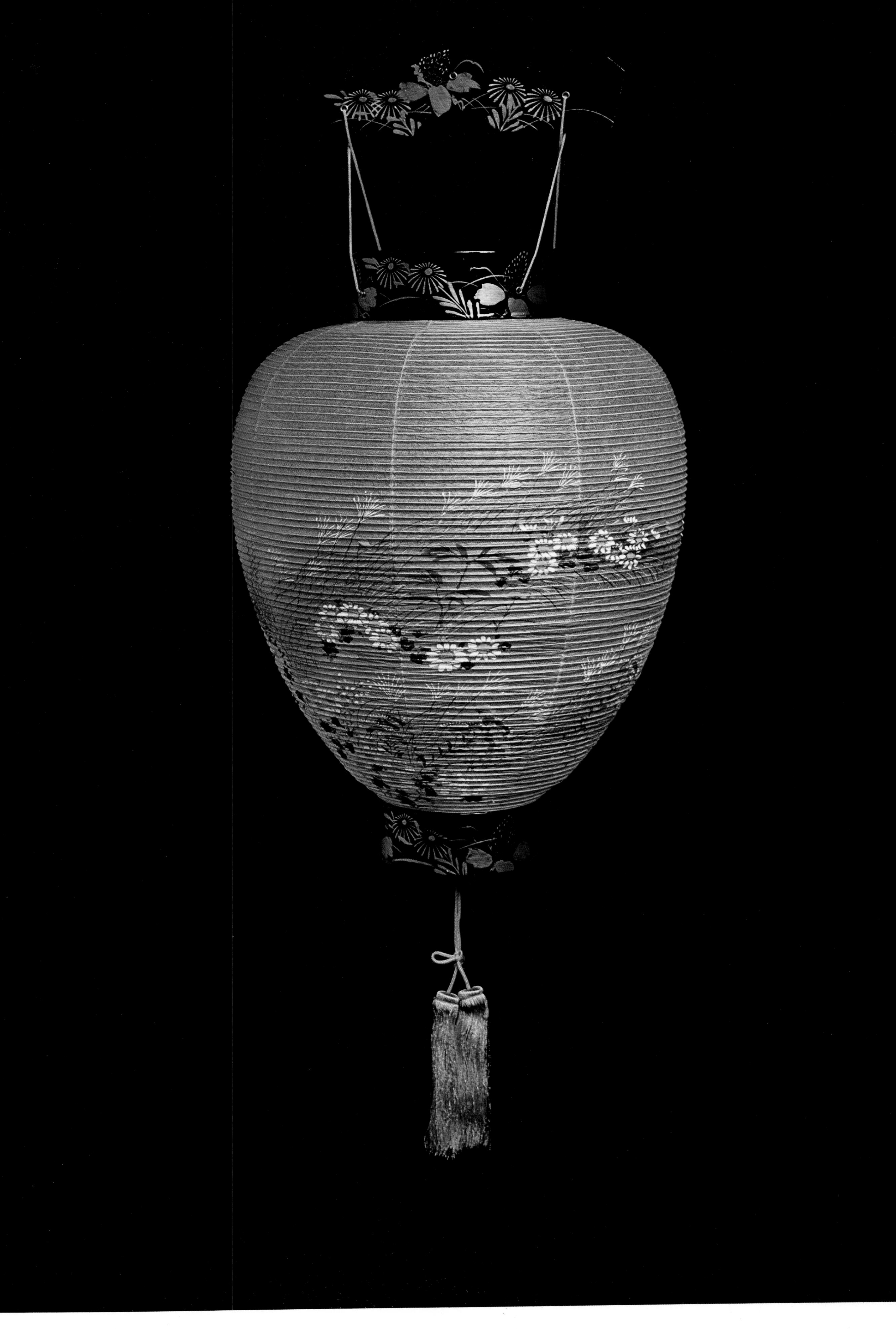

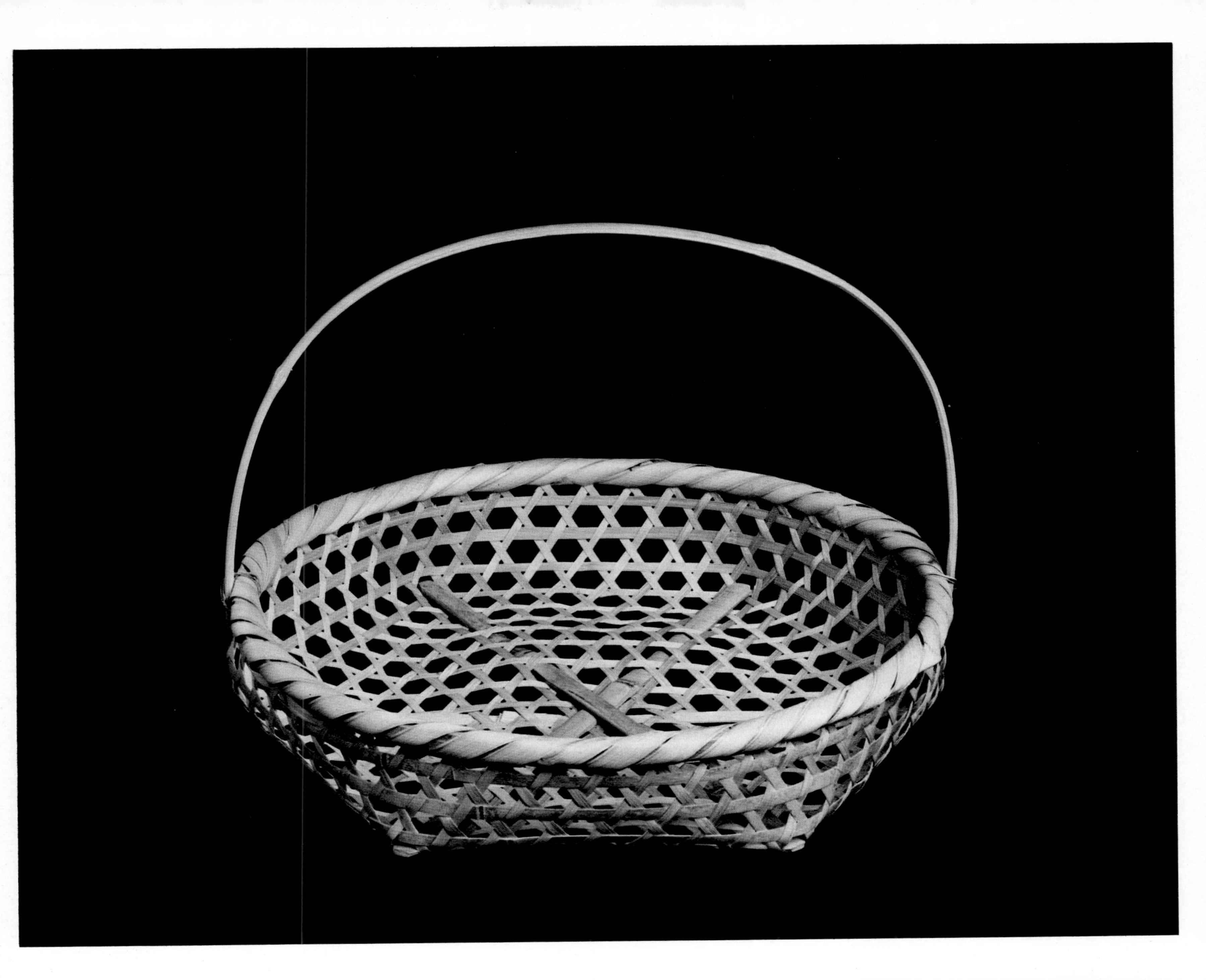

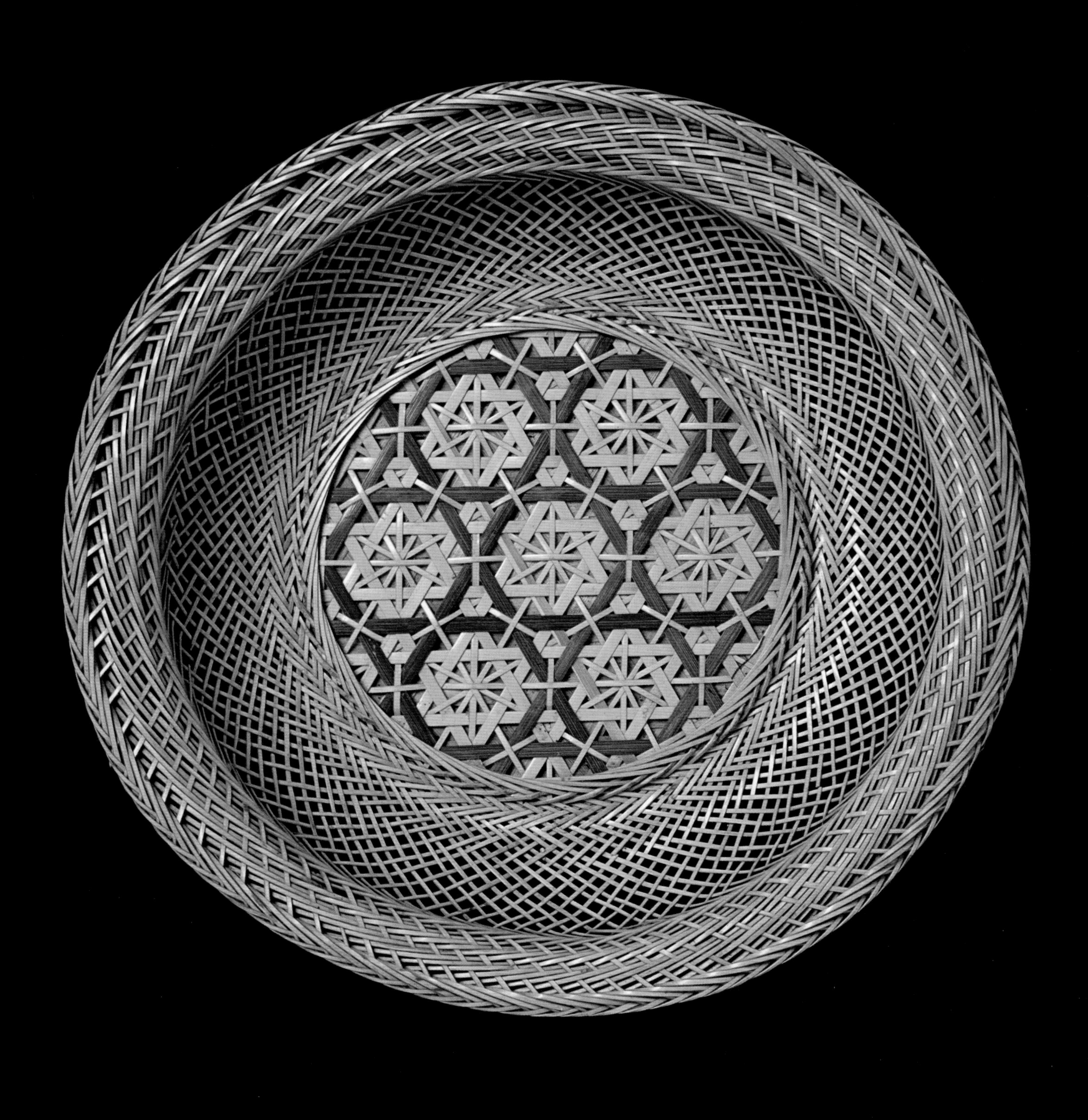

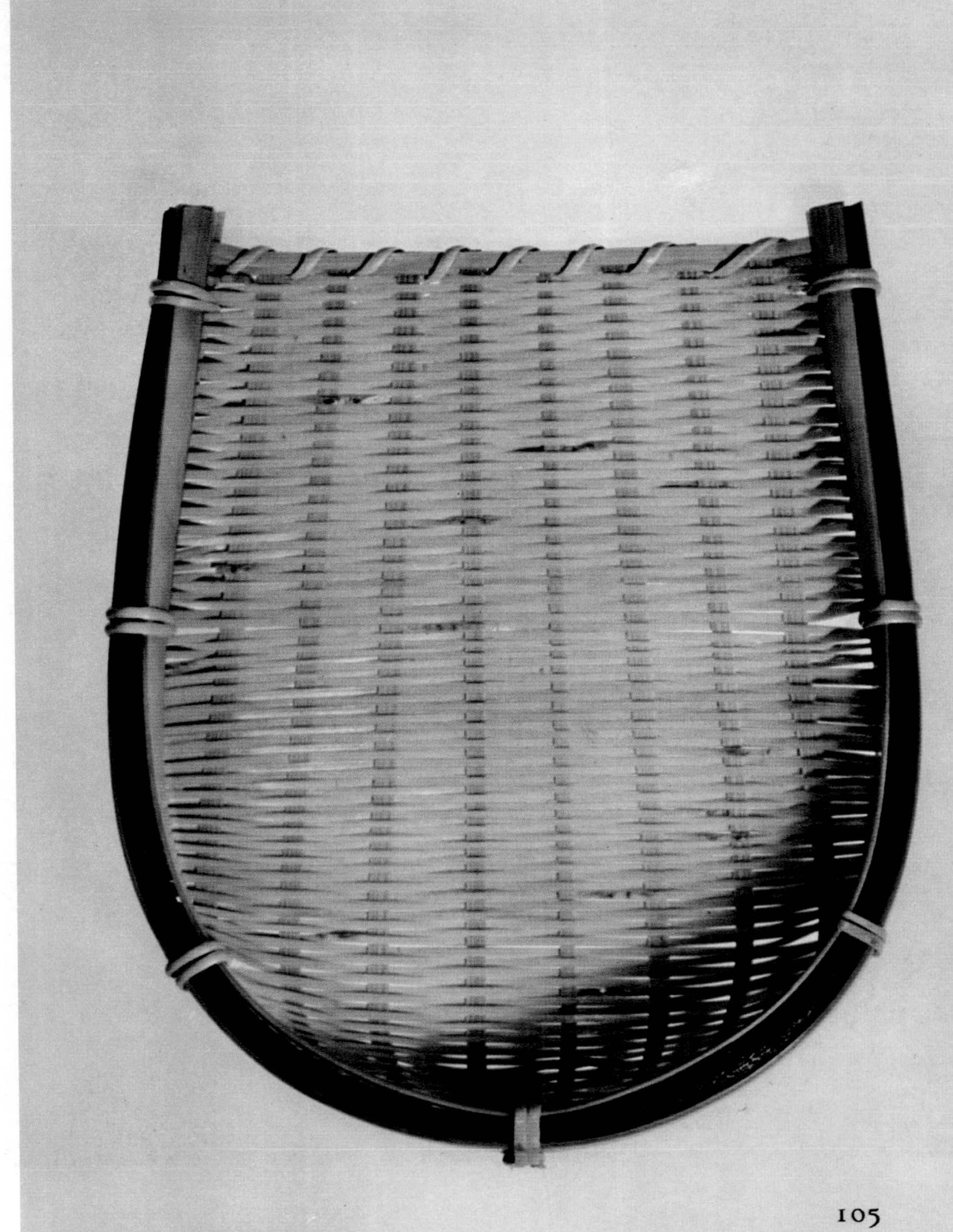

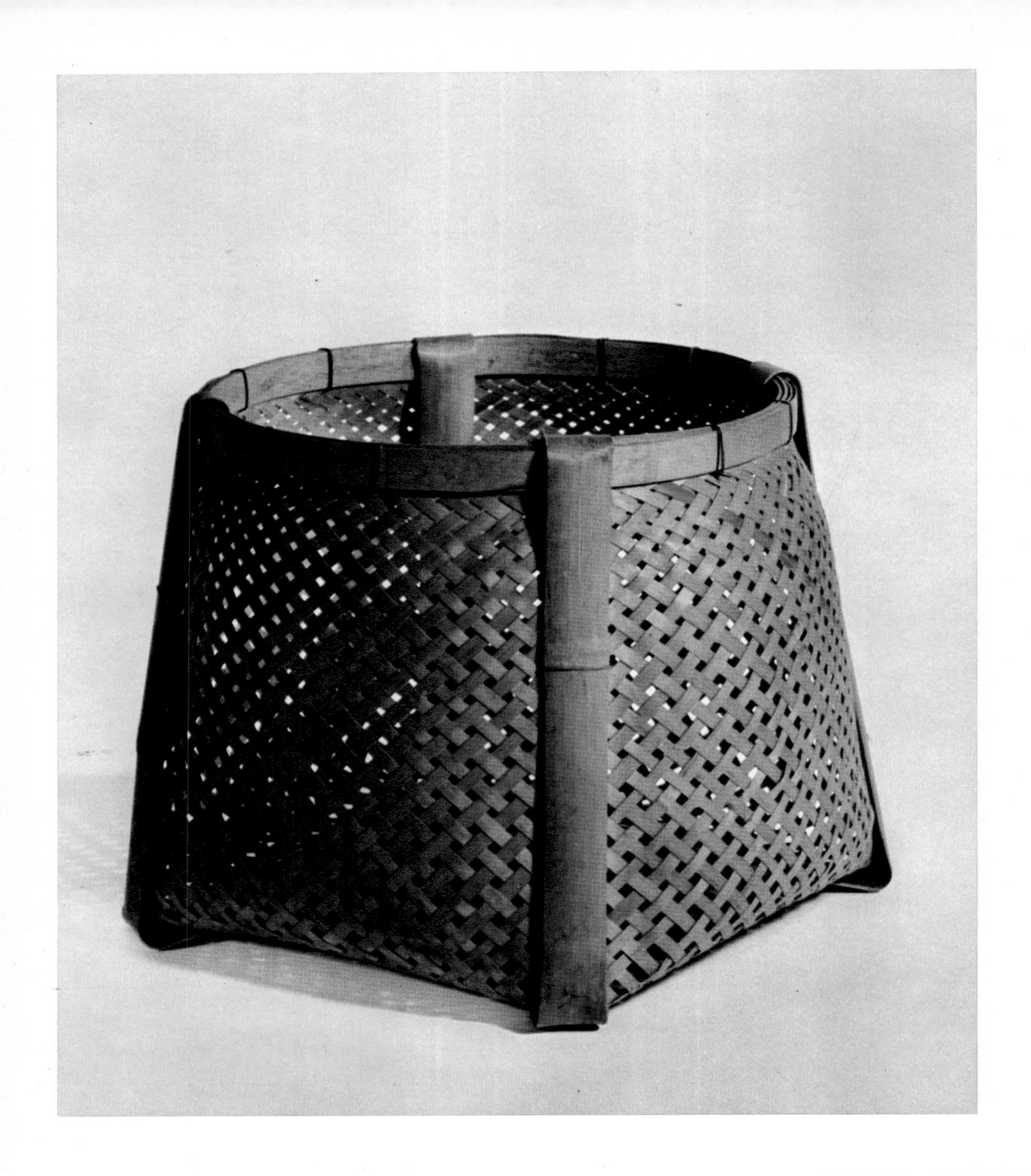

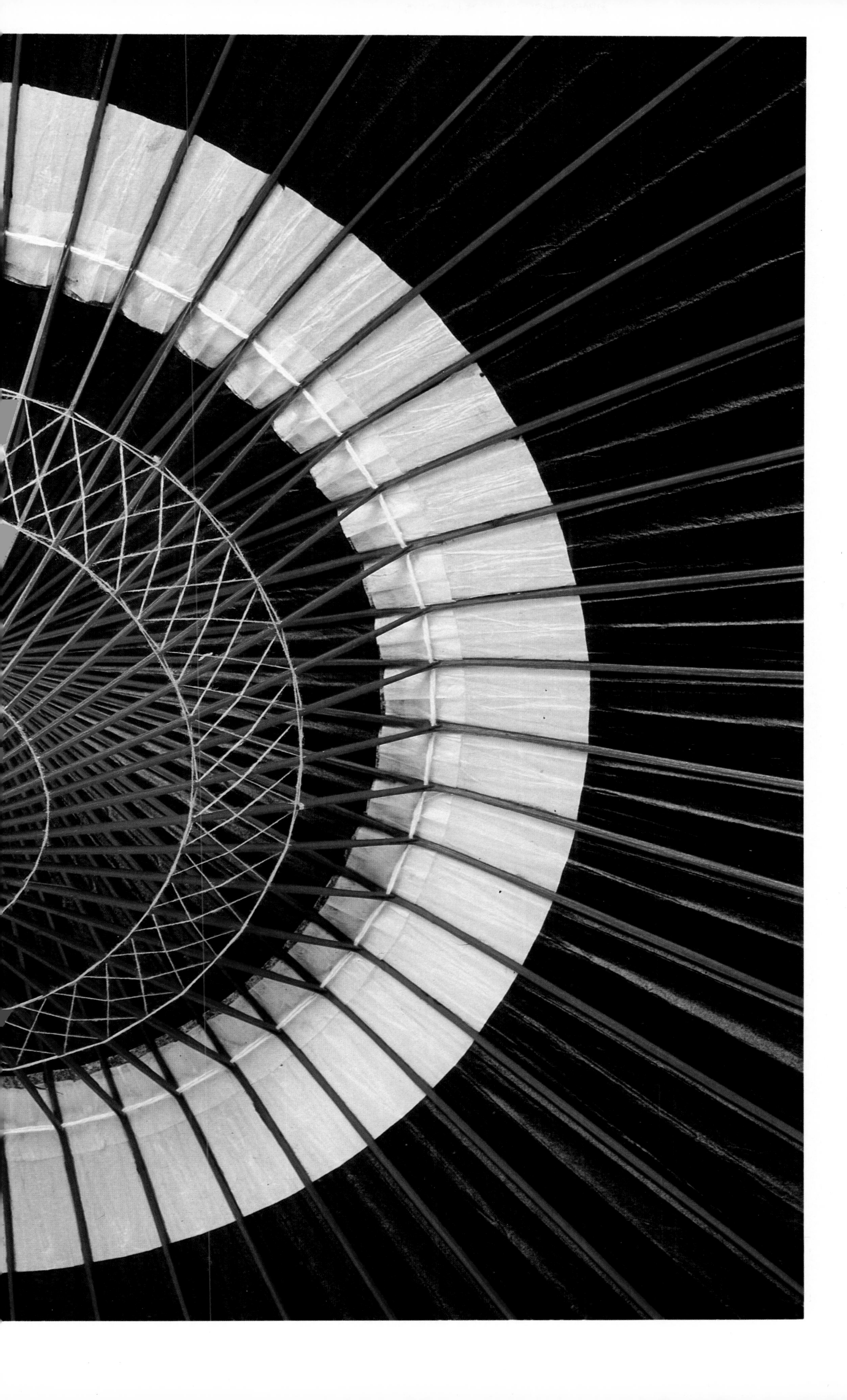

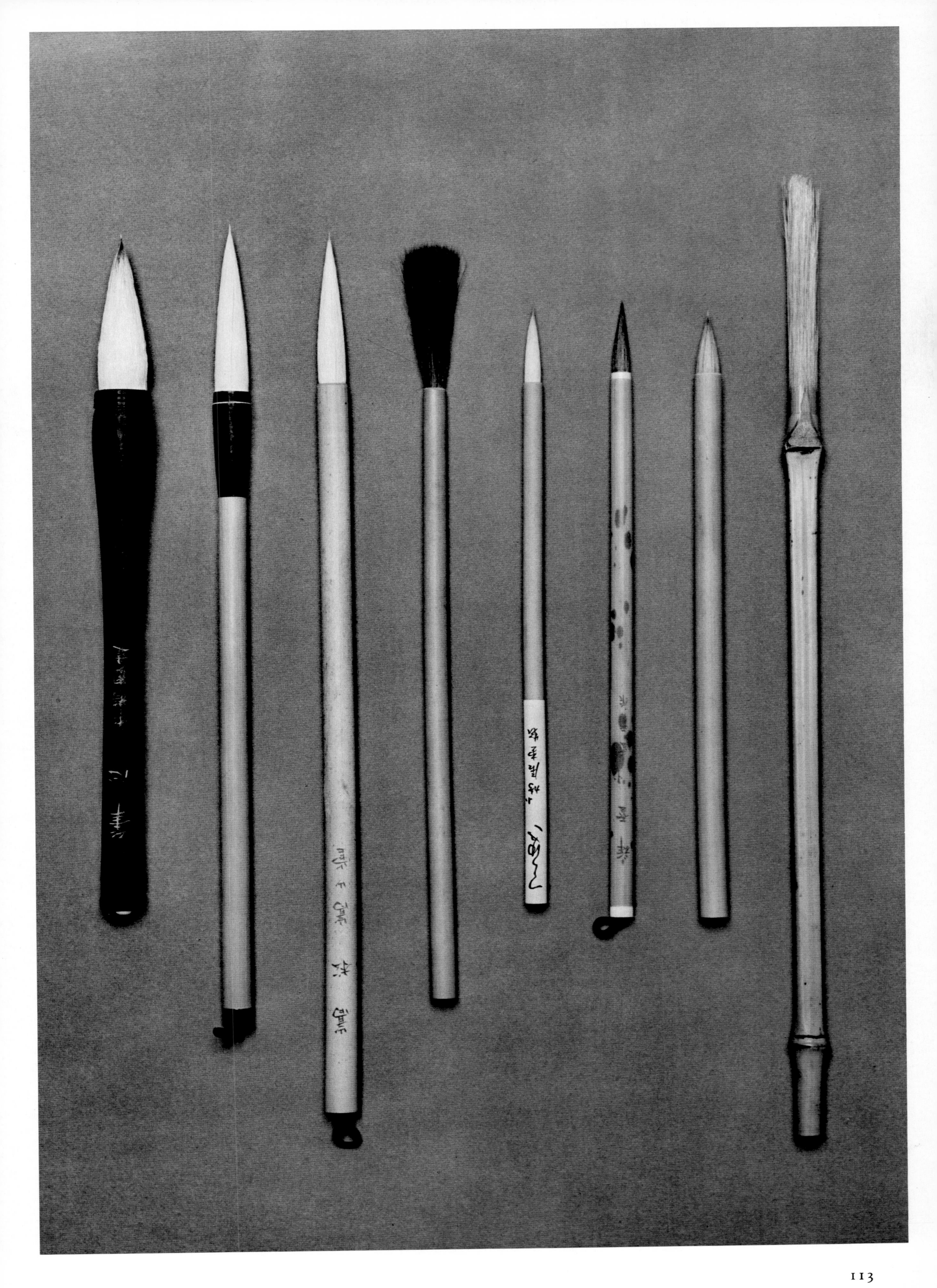

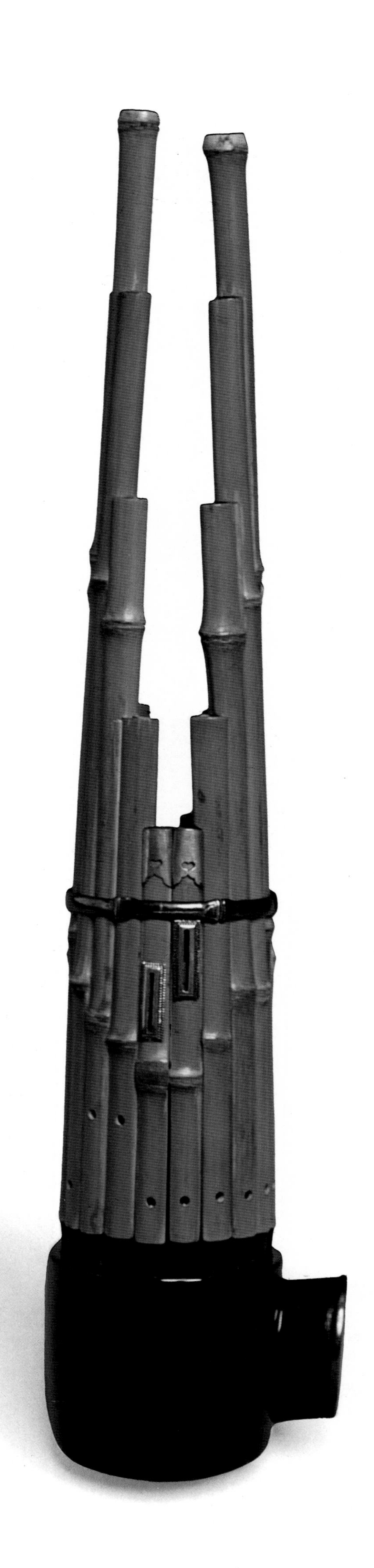

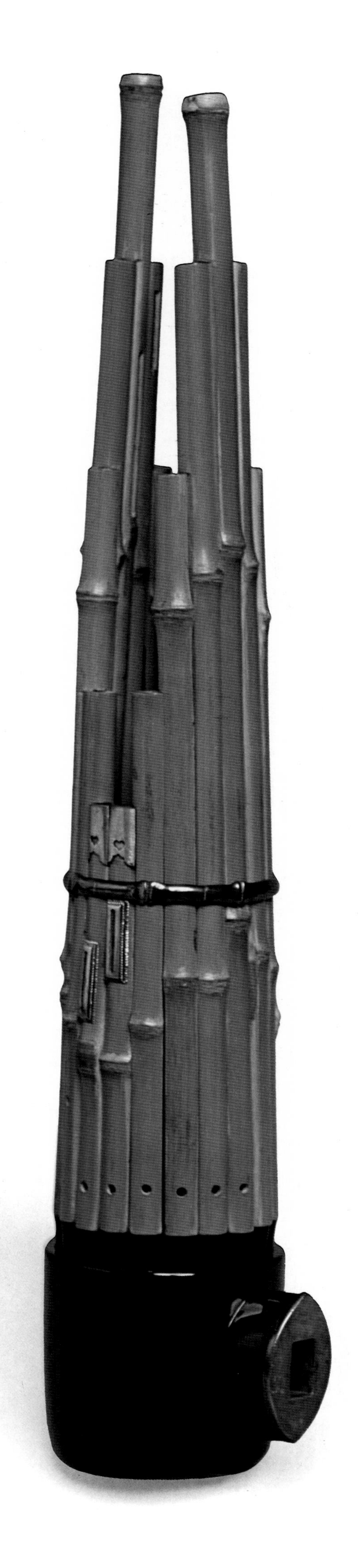

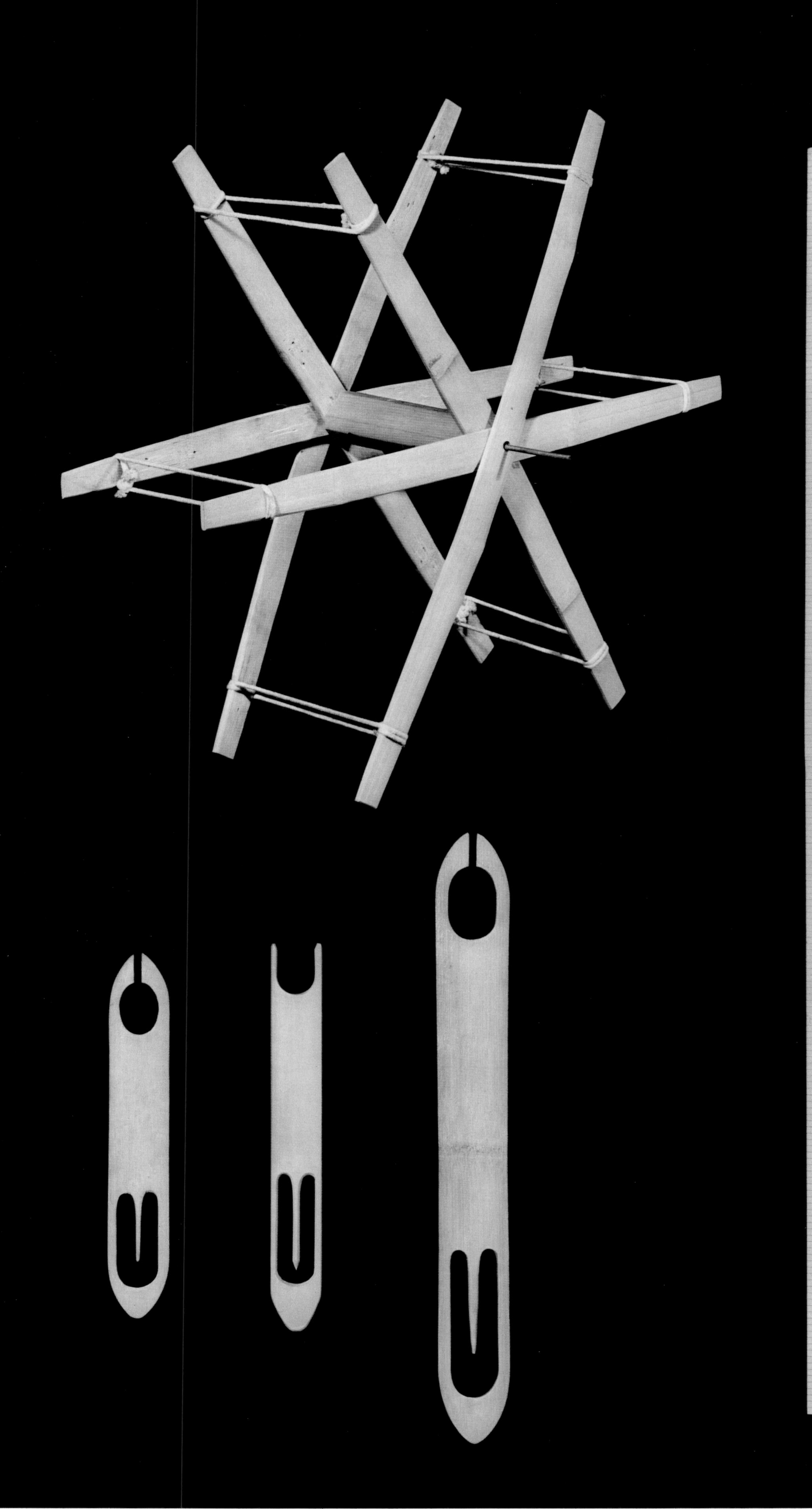

たから船

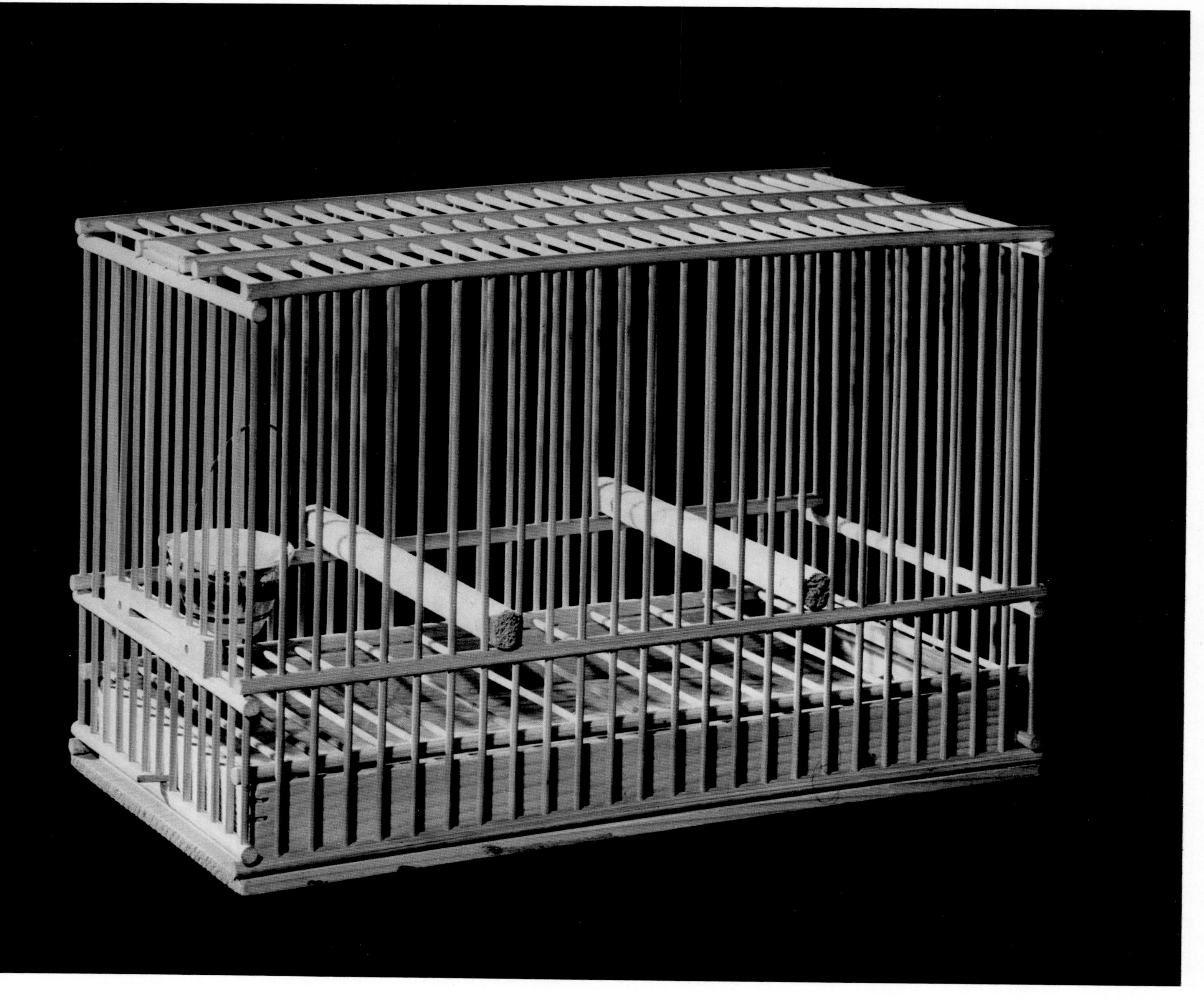

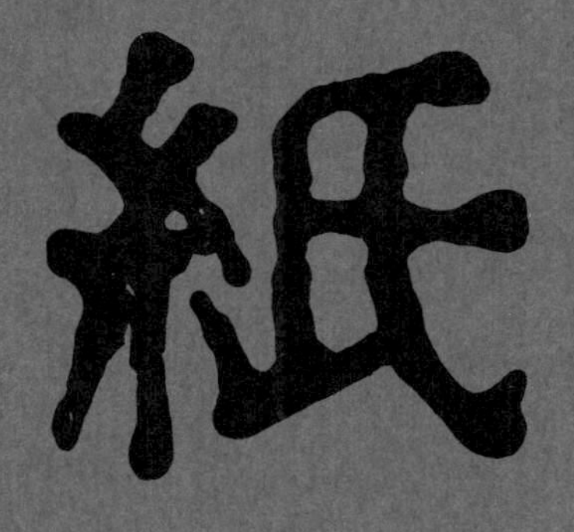

PAPER

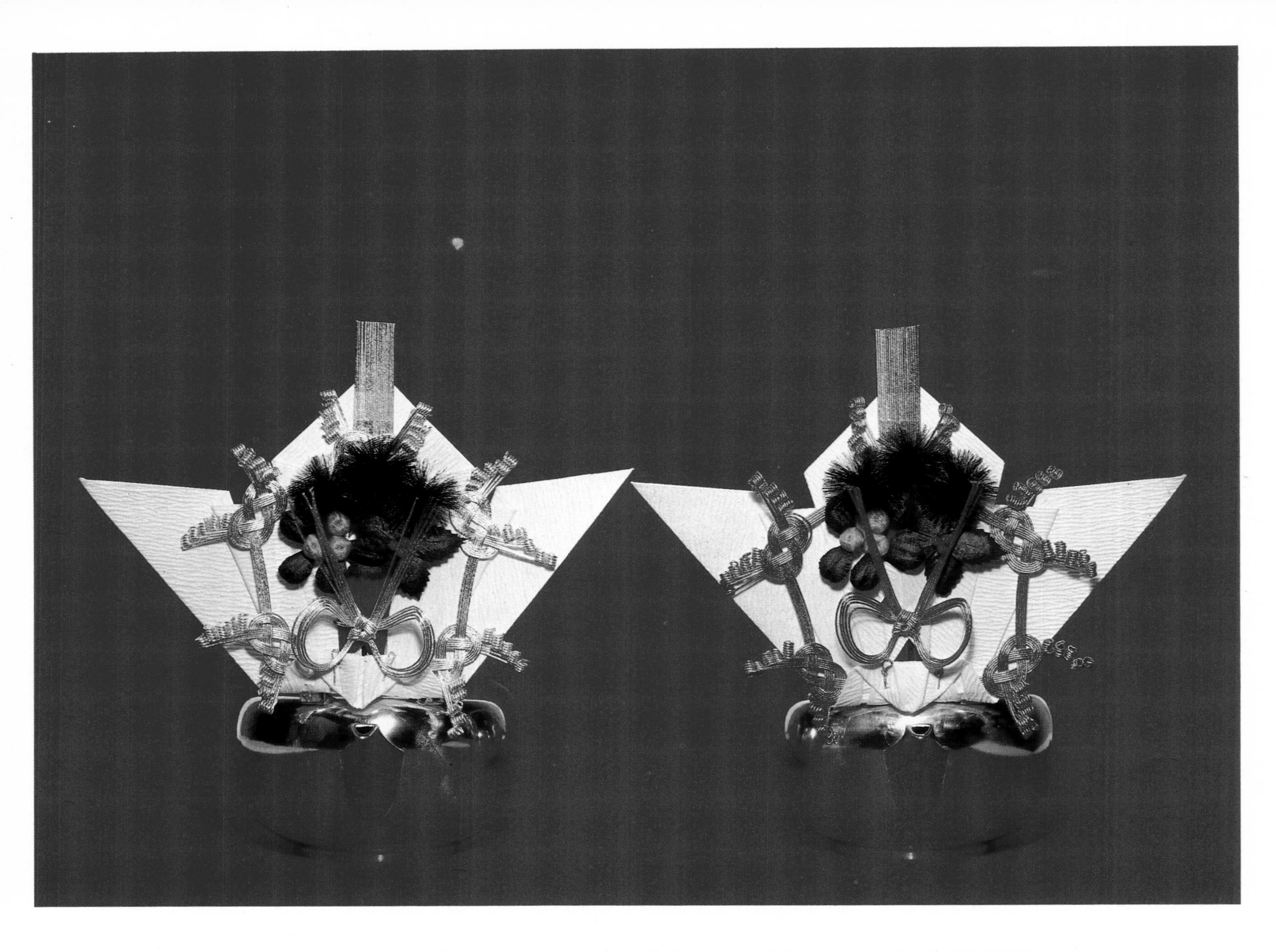

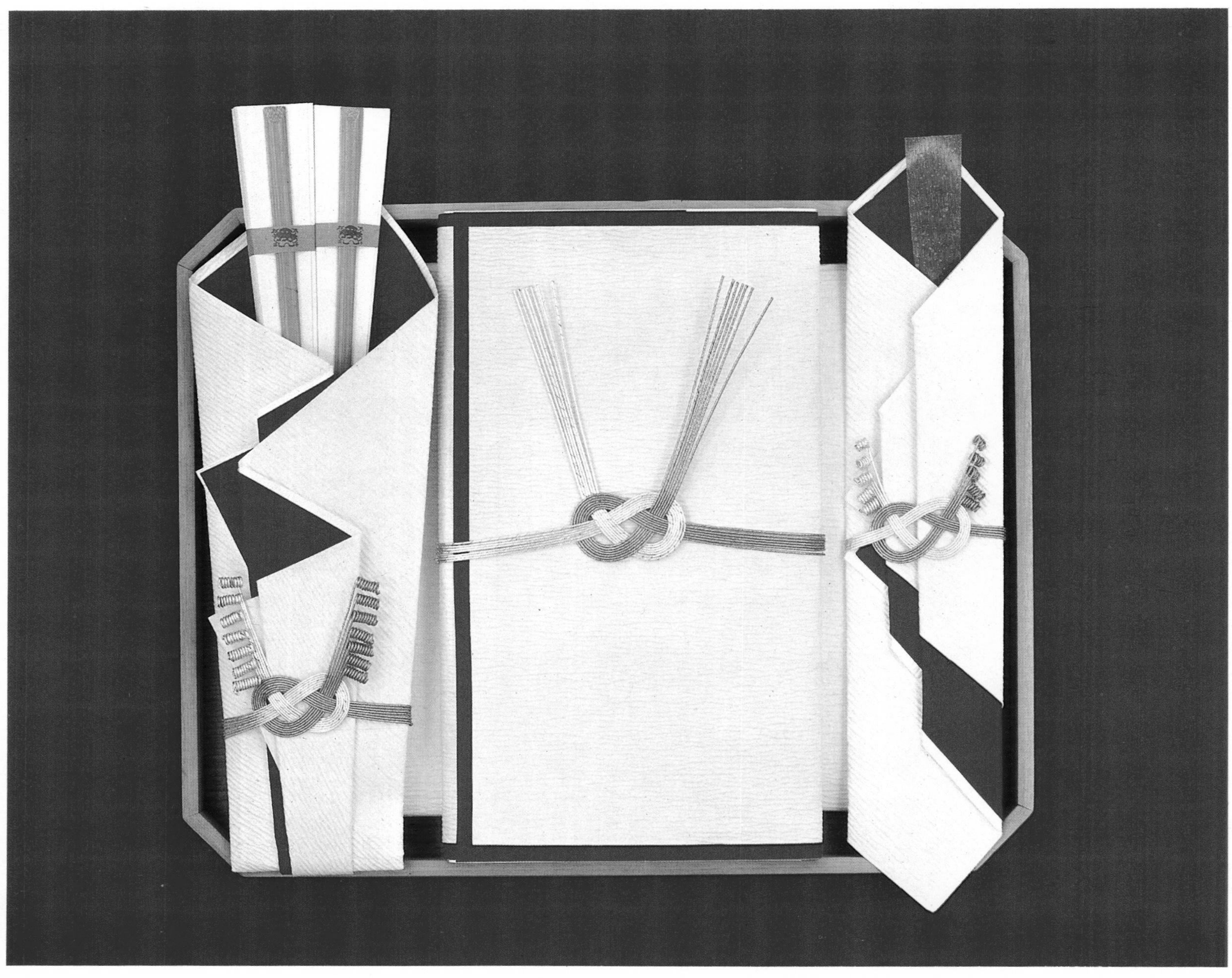

御祝

御祝儀

記念

目録

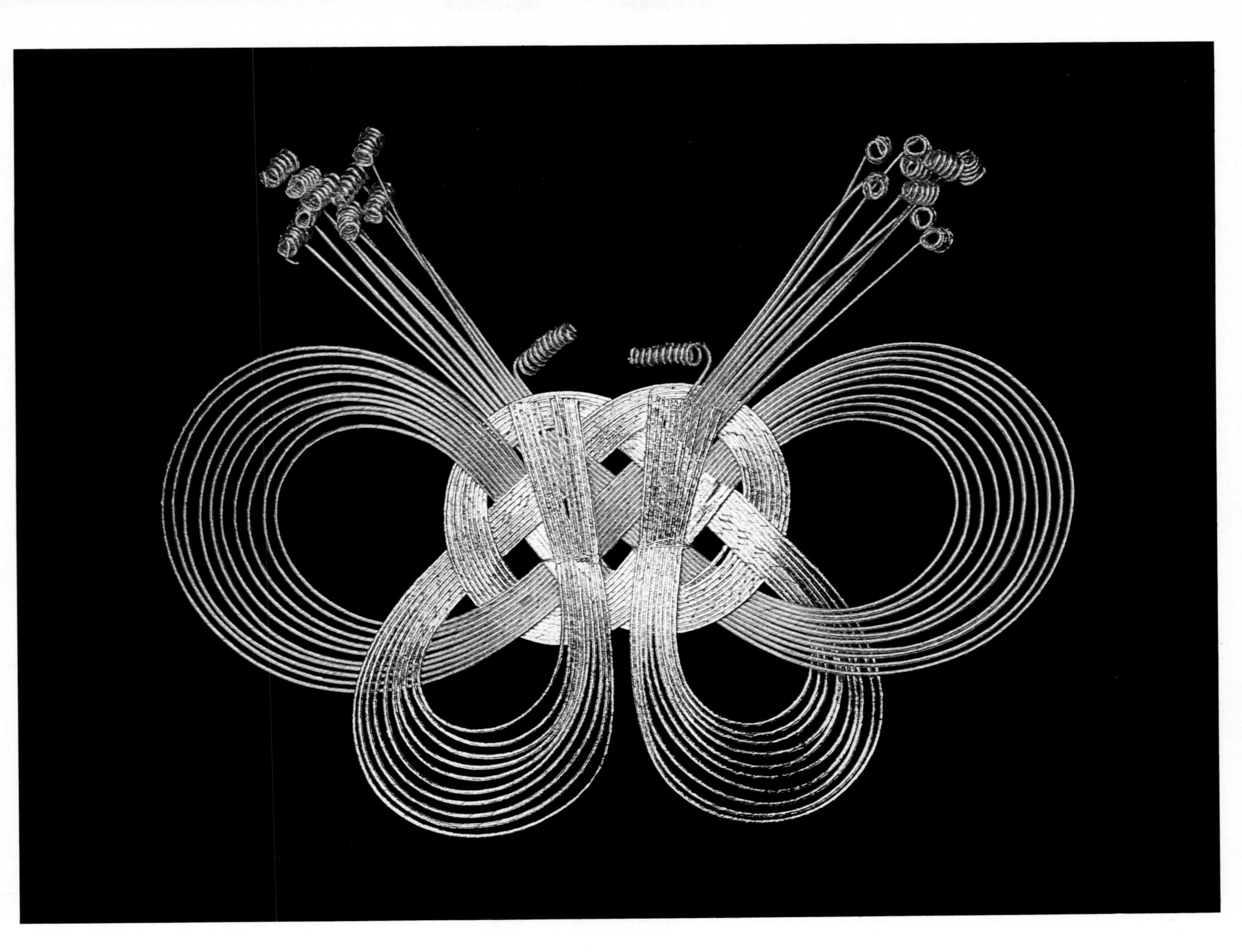

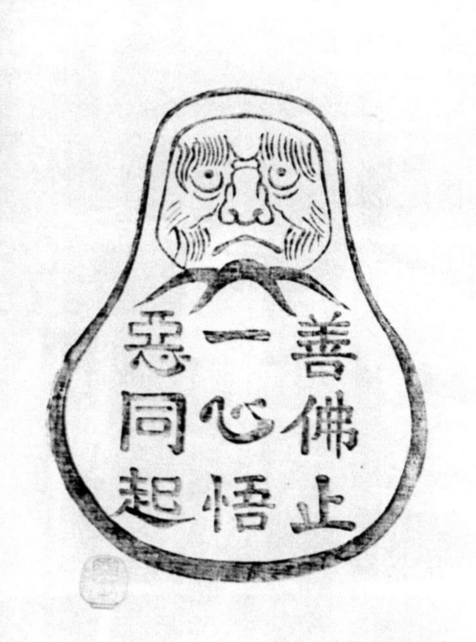
善佛上
一心悟
悪同起

面壁九年

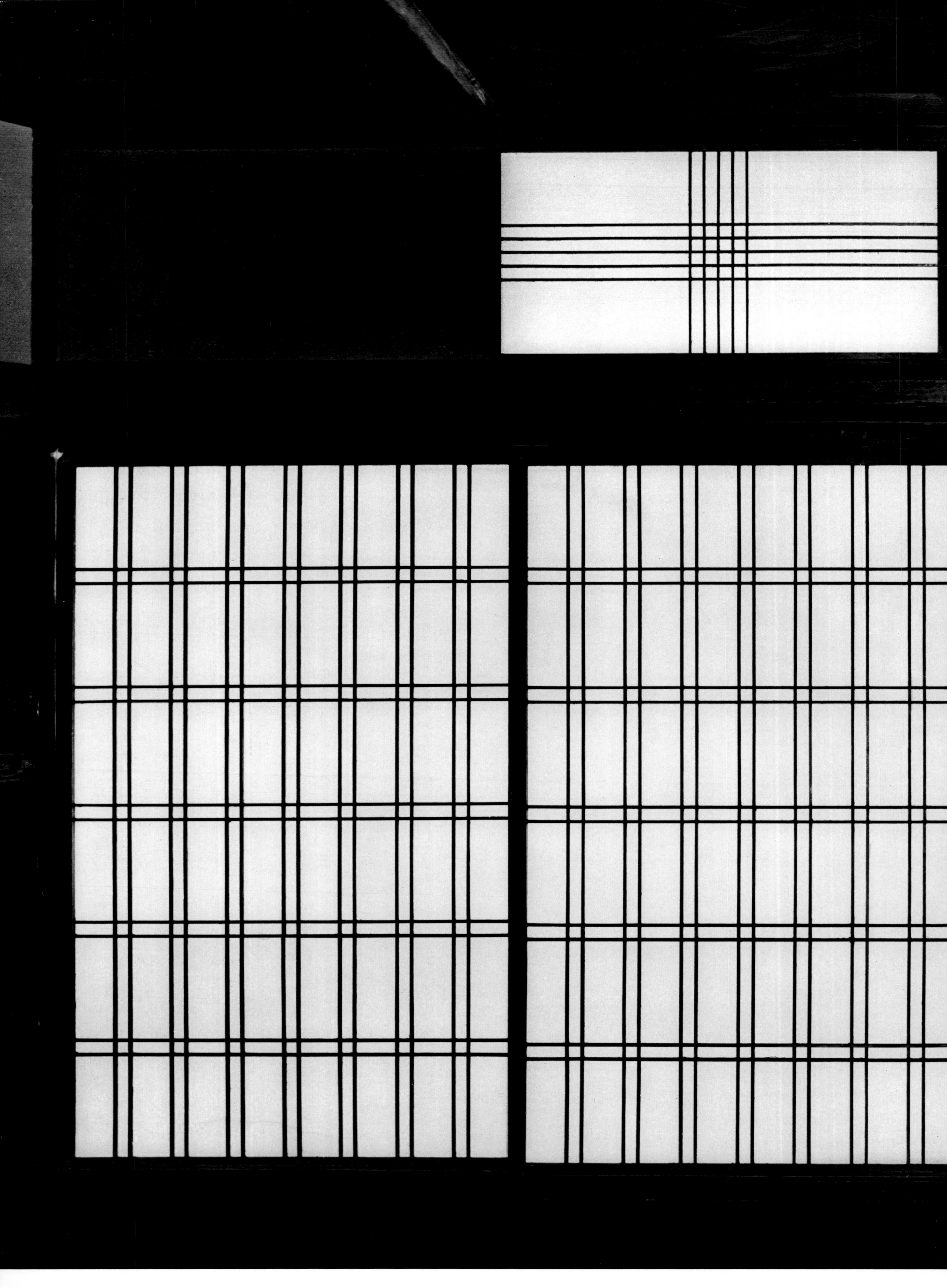

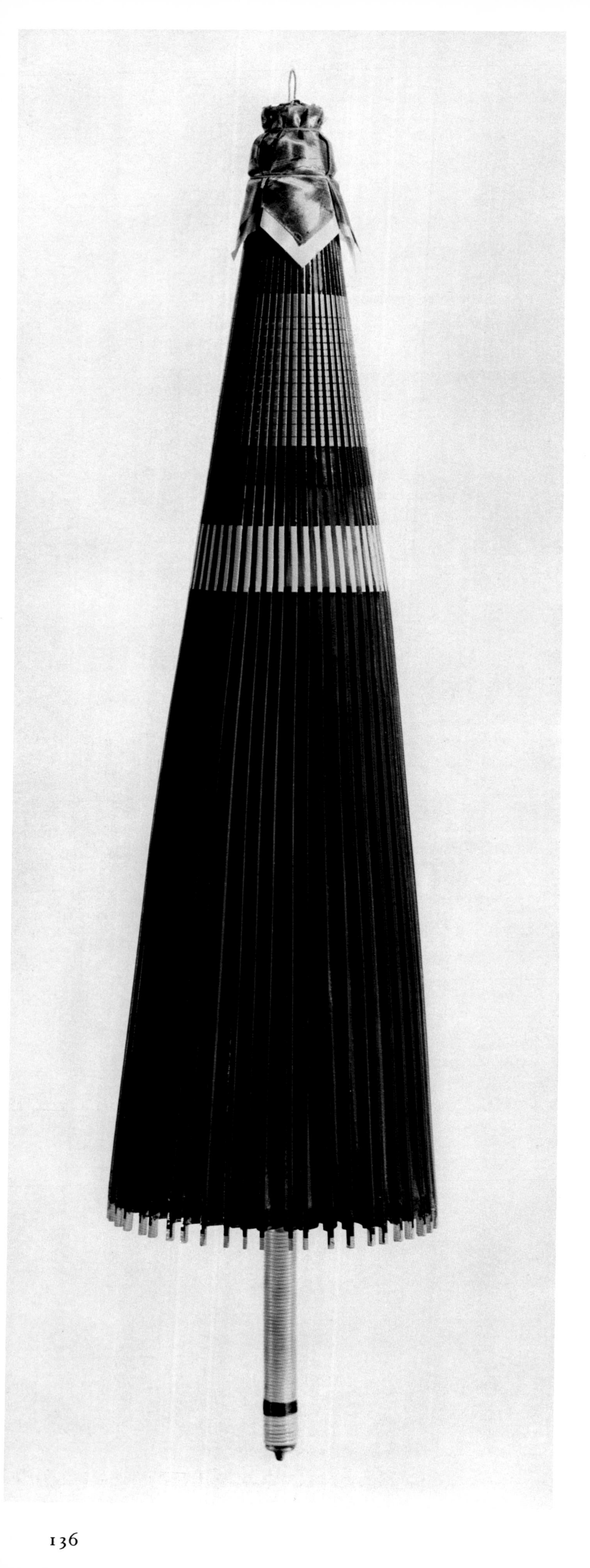

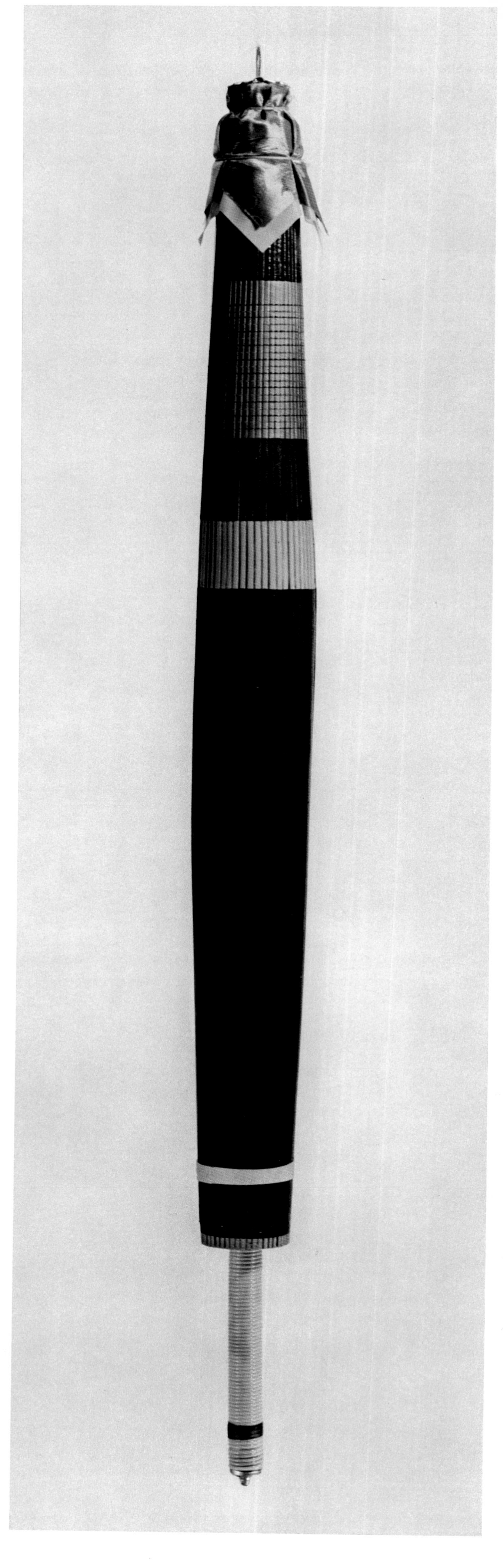

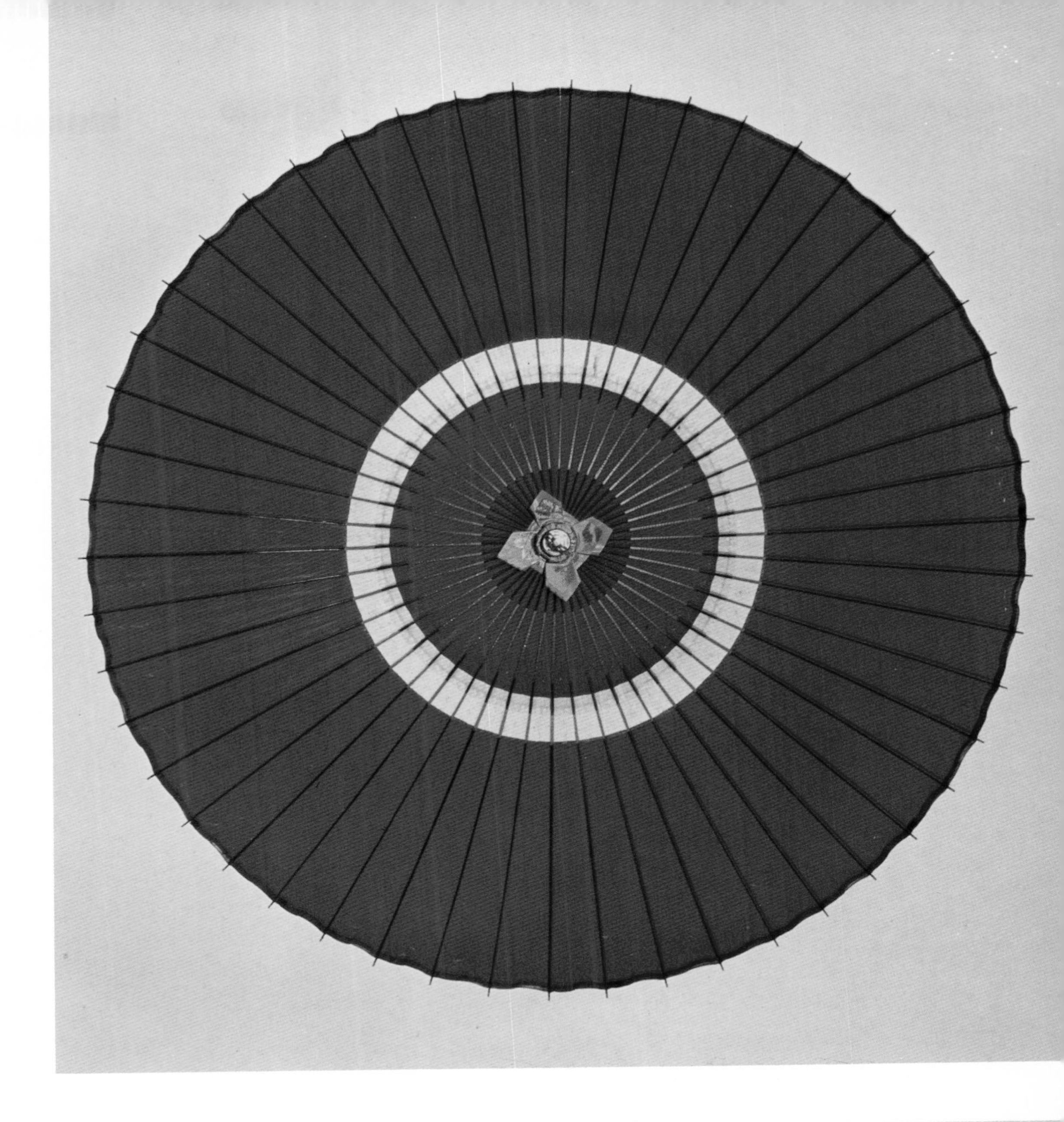

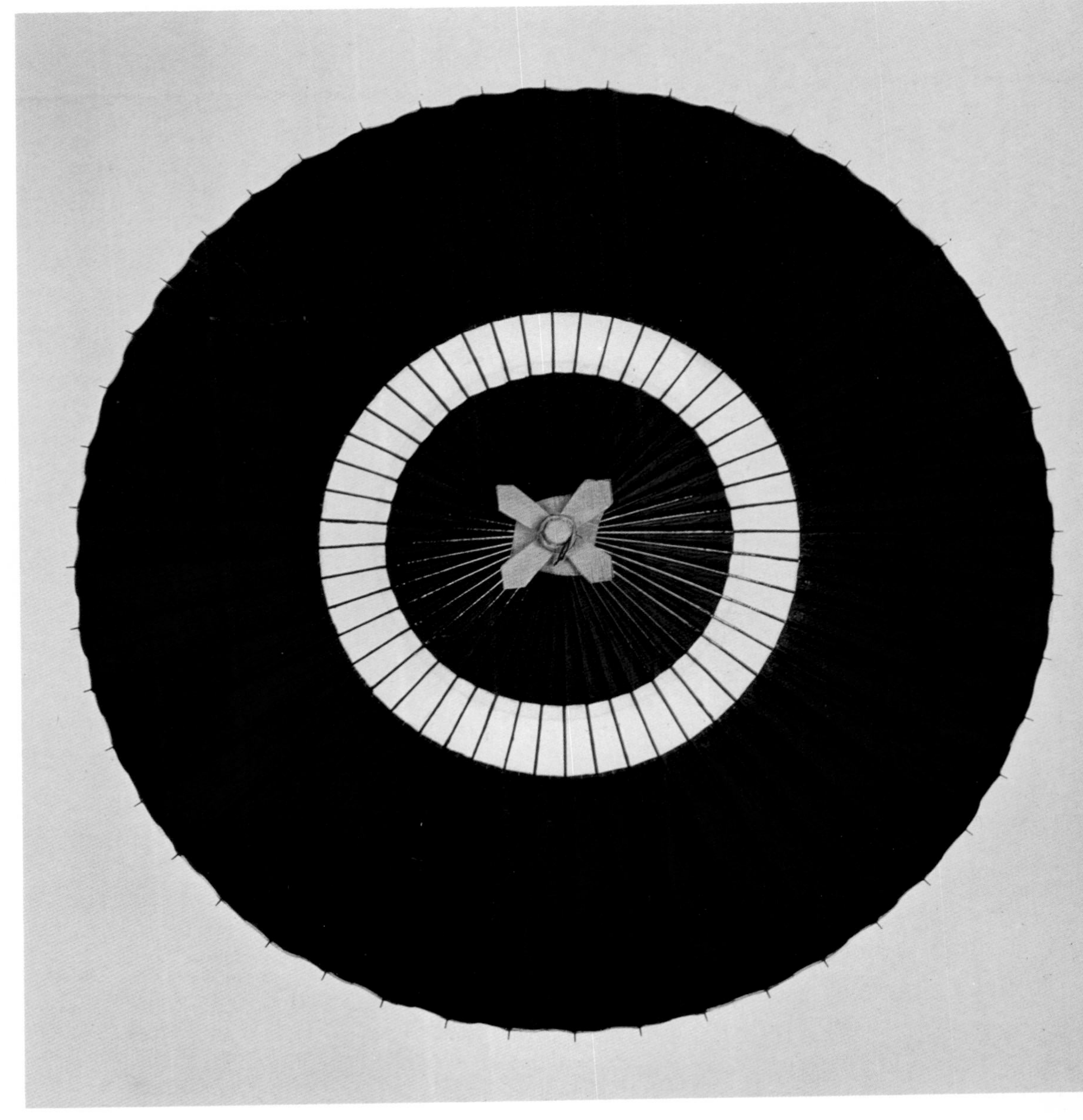

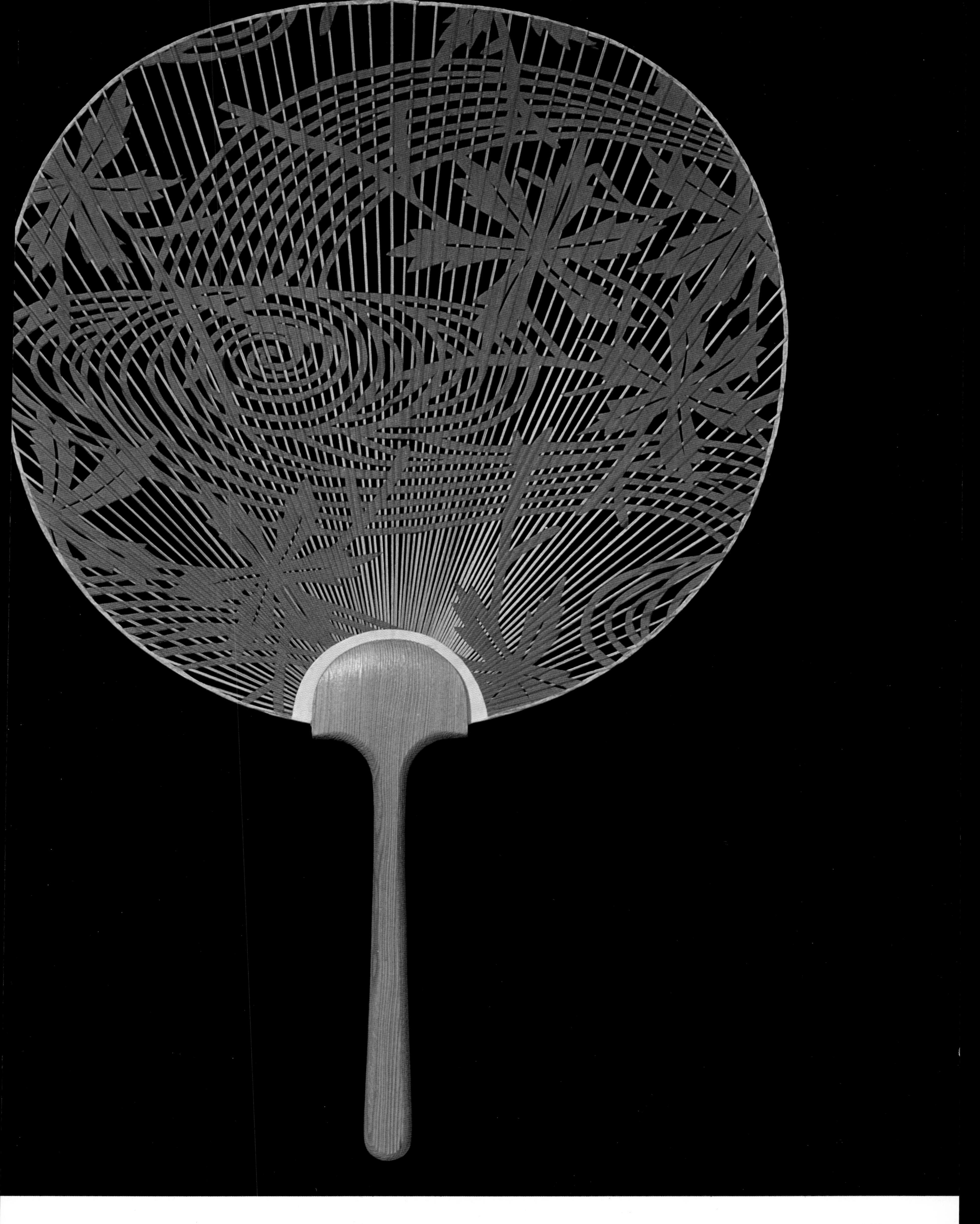

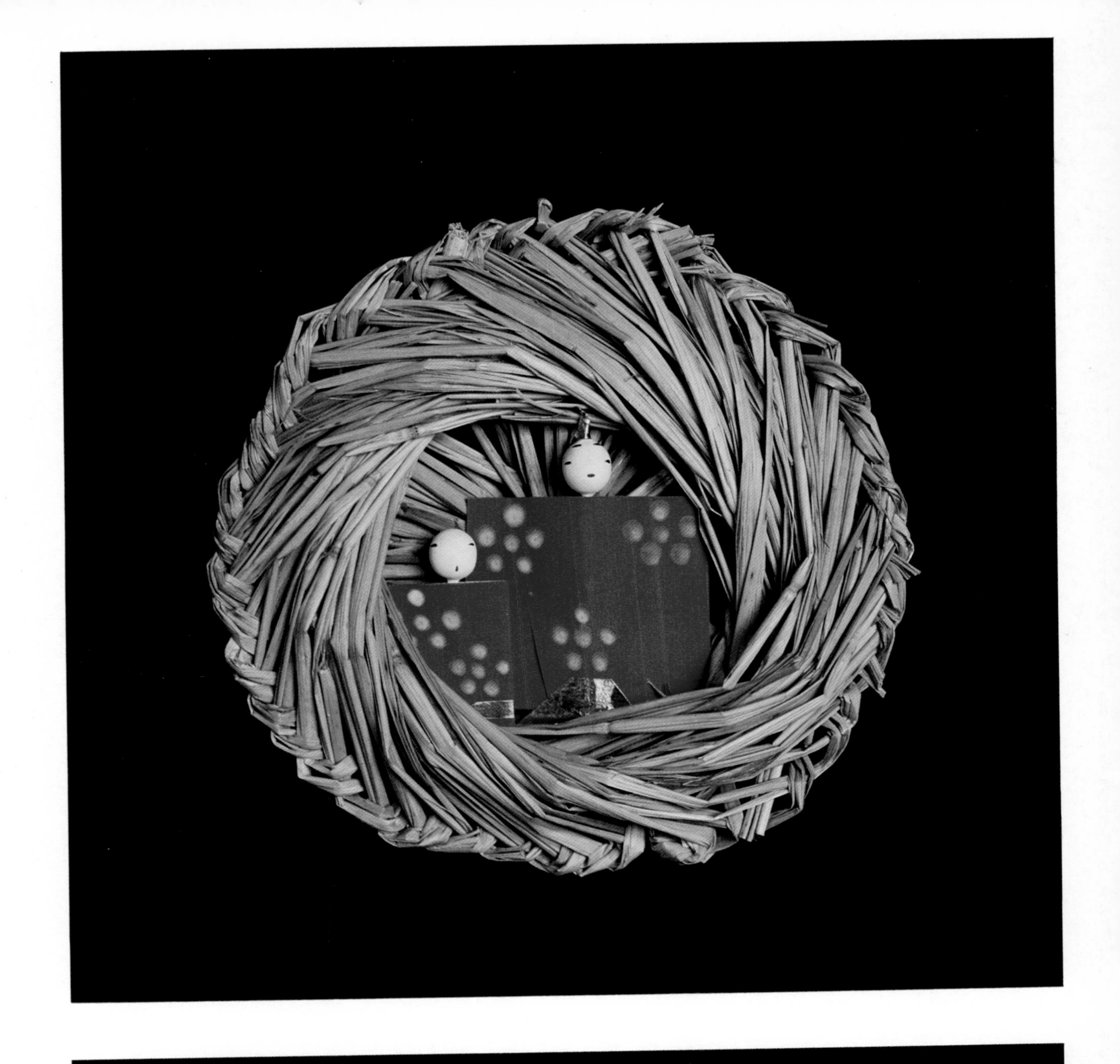

八坂皇大神

月鉾

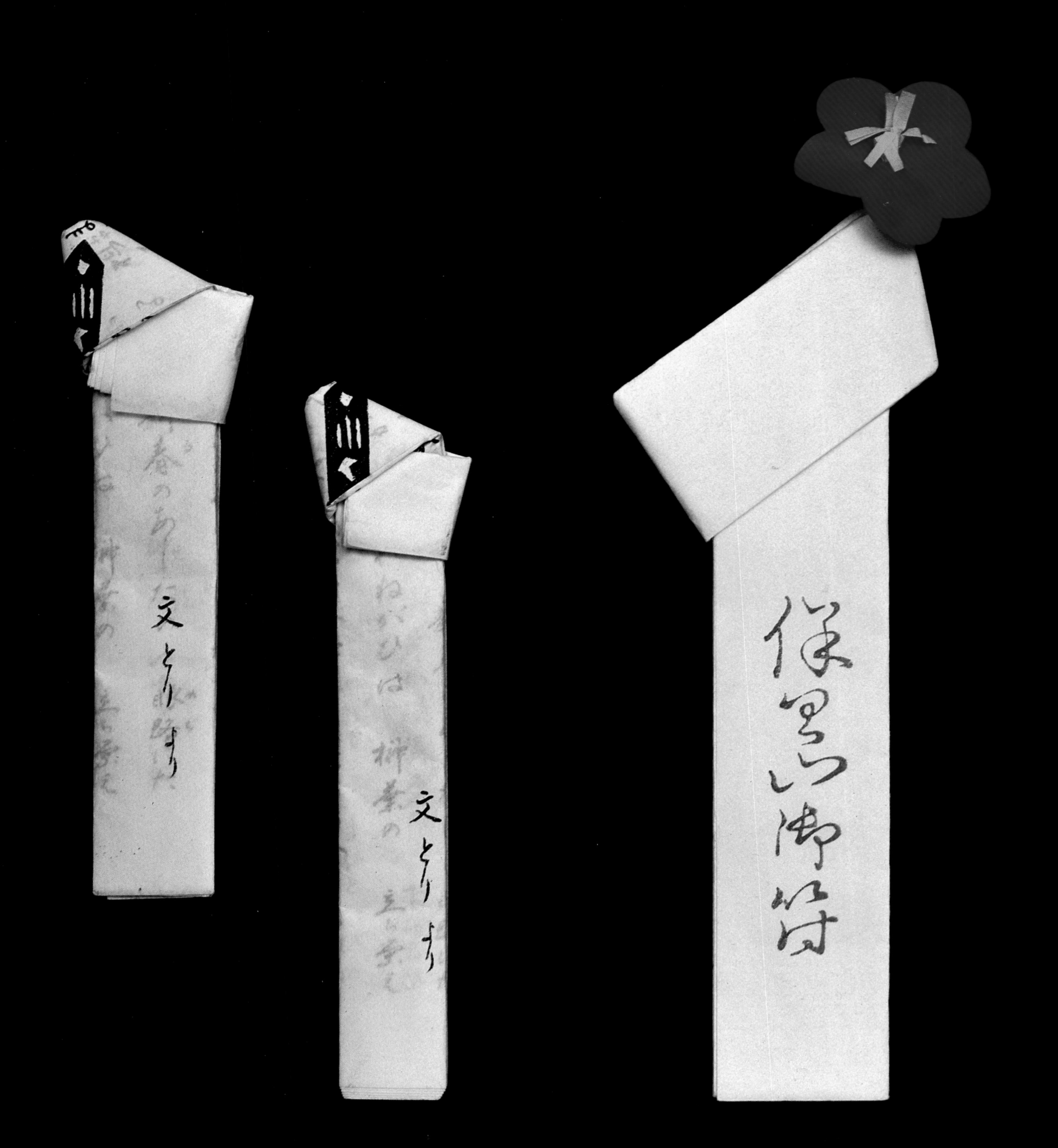
文とりより
文とりより

春日皇大神

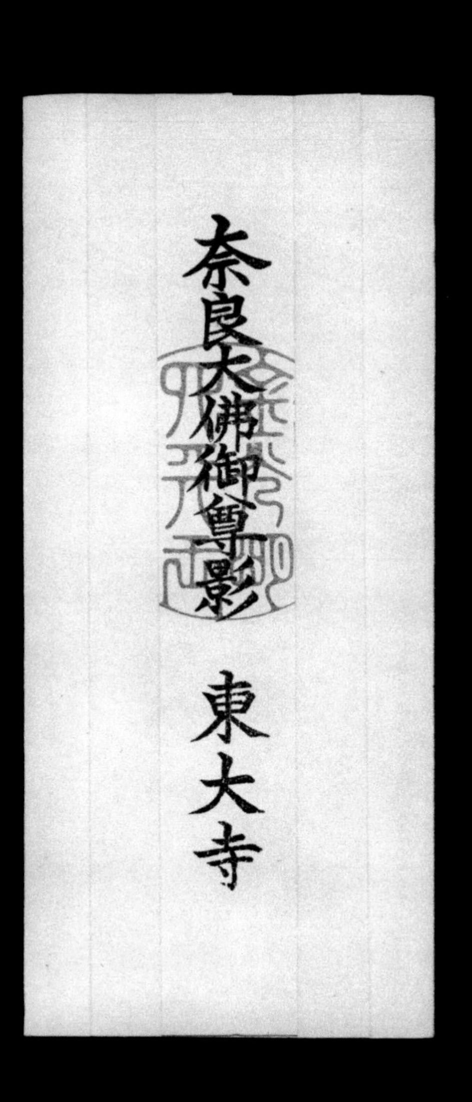
奈良大佛御尊影
東大寺

音羽山
千手観音供守護
清水寺

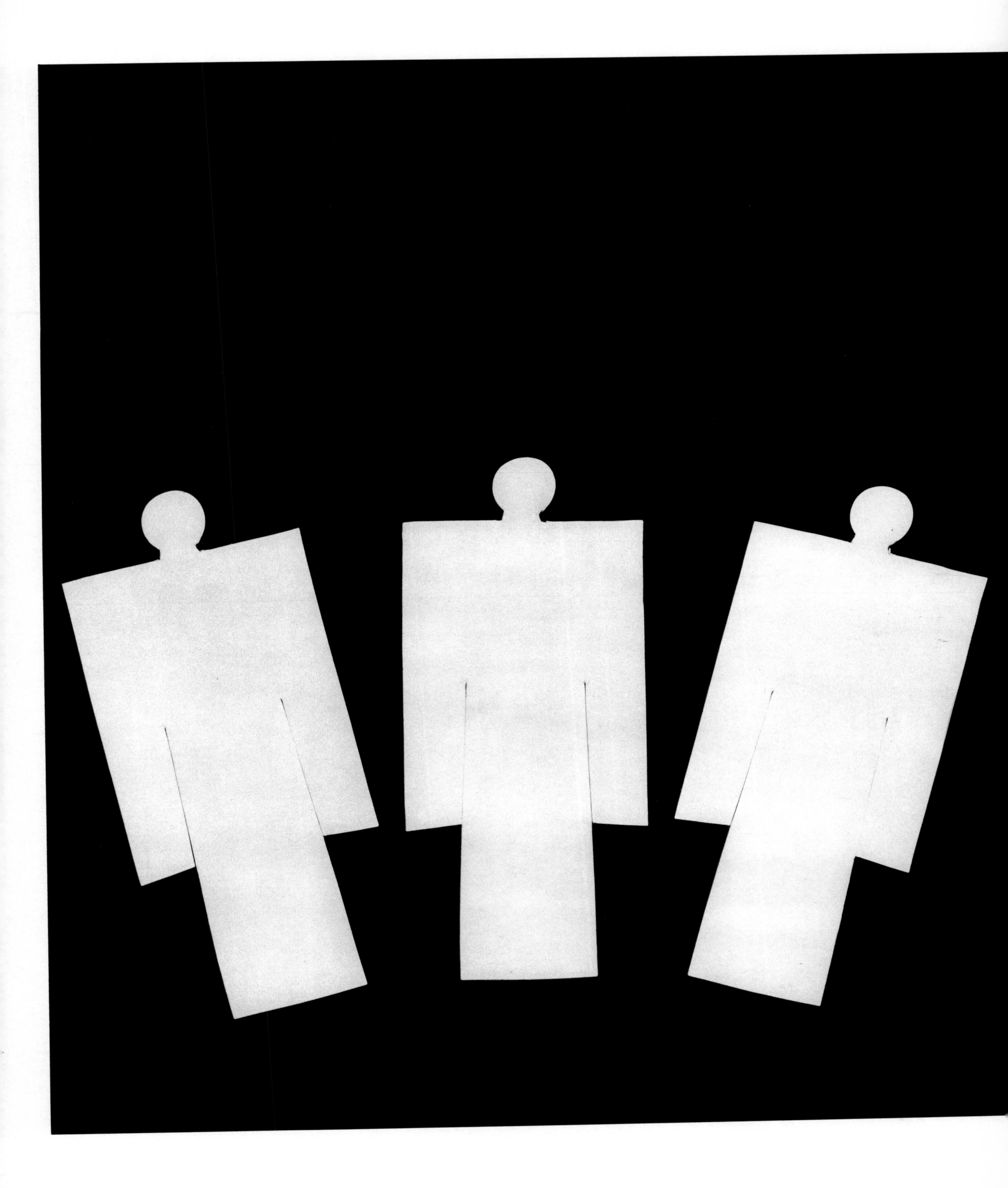

住吉神社守護
稲荷大神守護所

第十三大吉
清三寶荒神王御籤
手把太陽輝
東君發舊枝
稼苗方欲秀
猶更上雲梯
第六十四凶
清三寶荒神王御籤
安居且慮危
第四十七吉
清三寶荒神王御籤
更望身前立
何期在晩成
若逢重山去
好縁自相迎

METAL

169

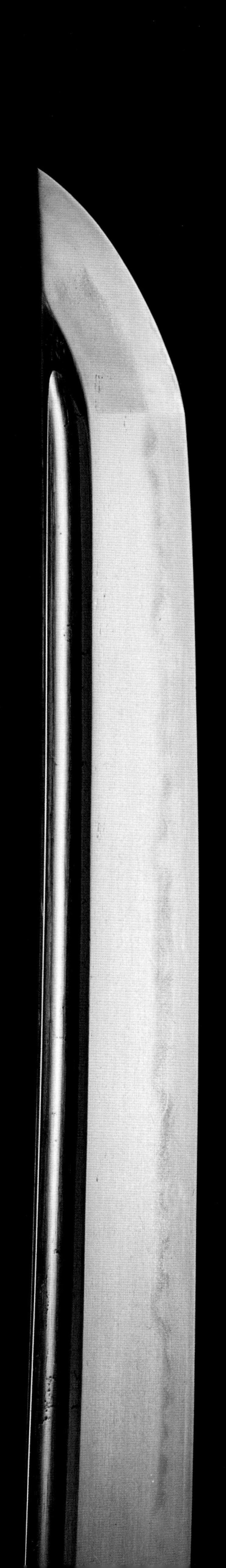

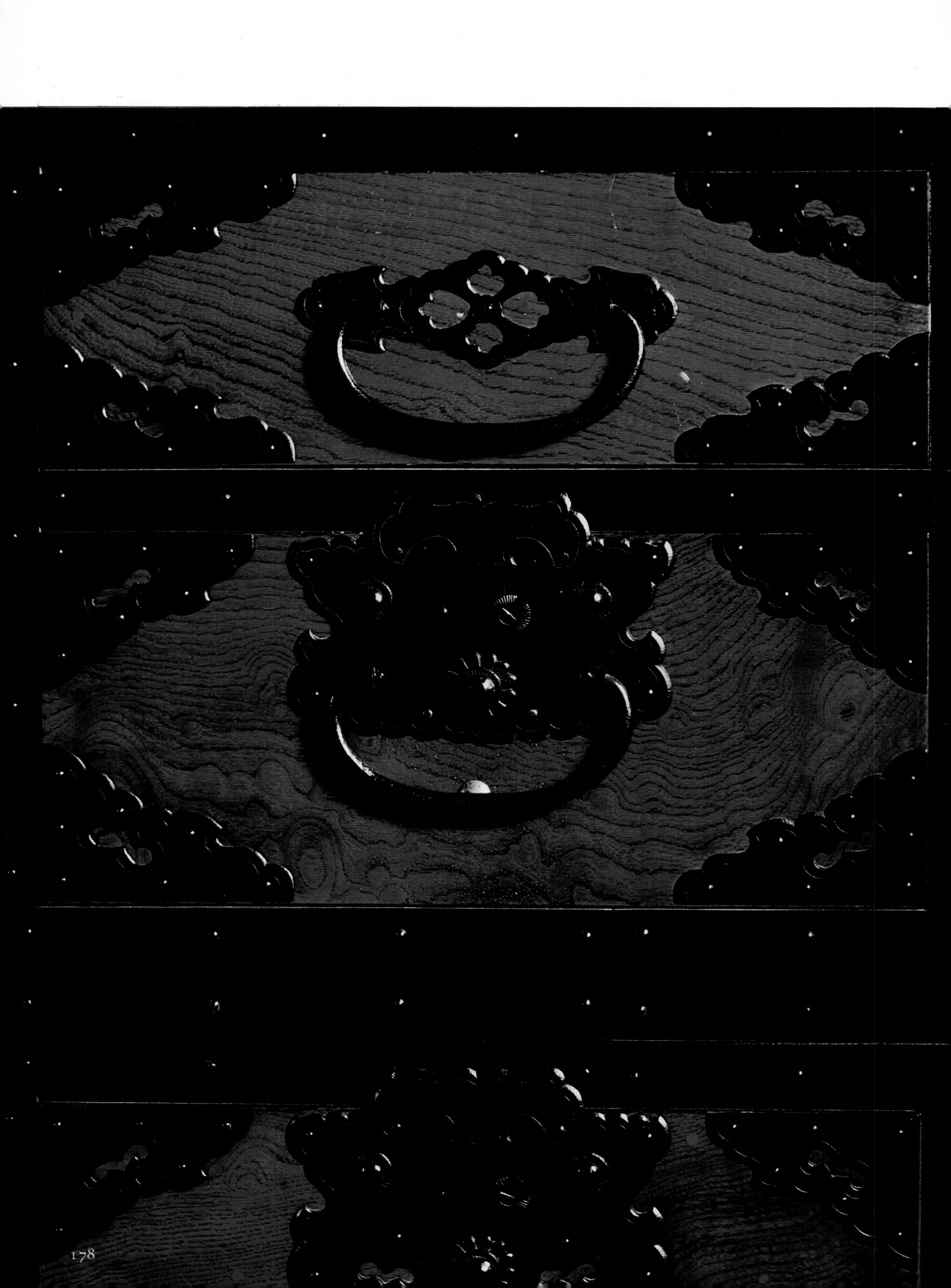

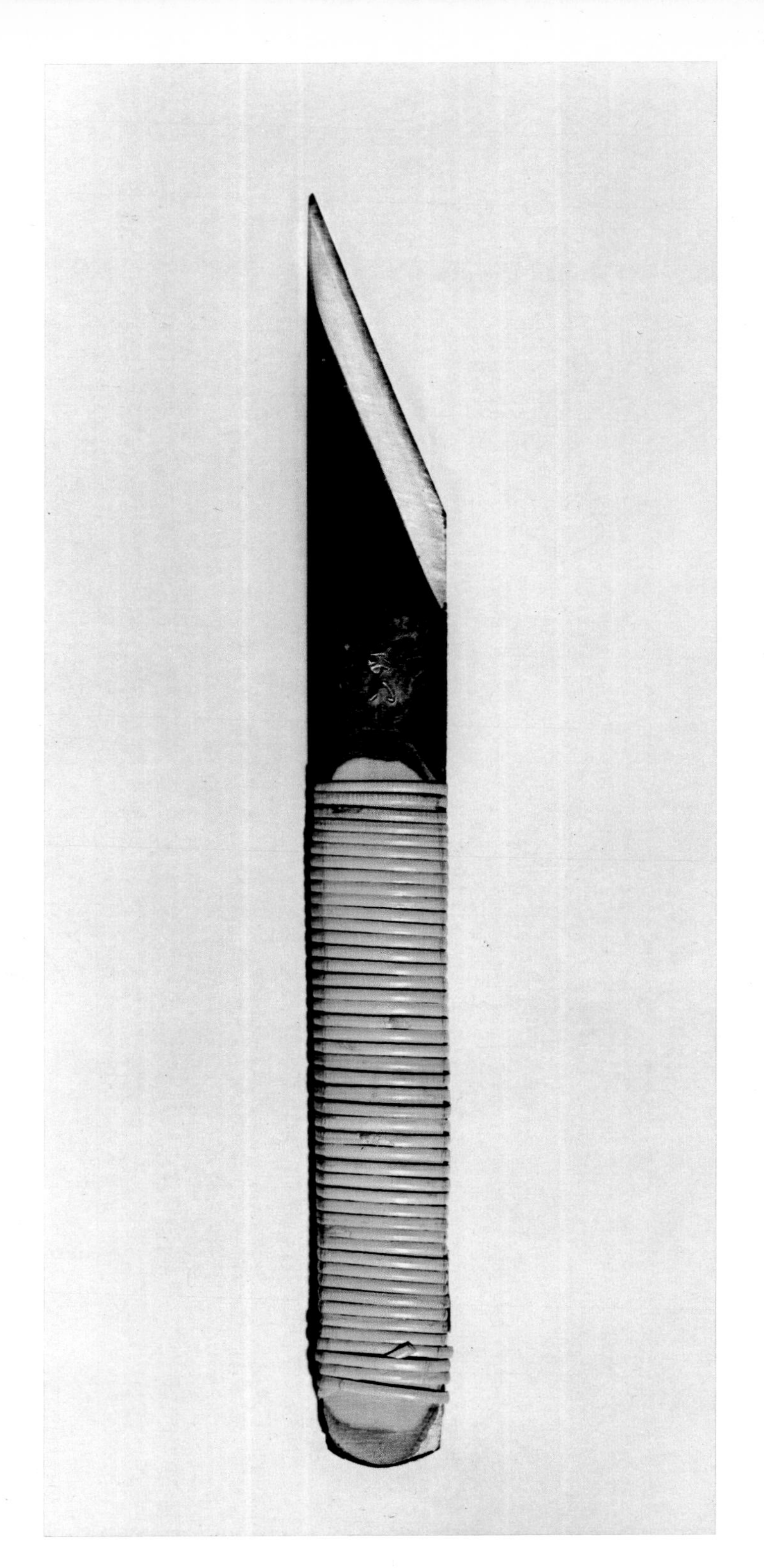

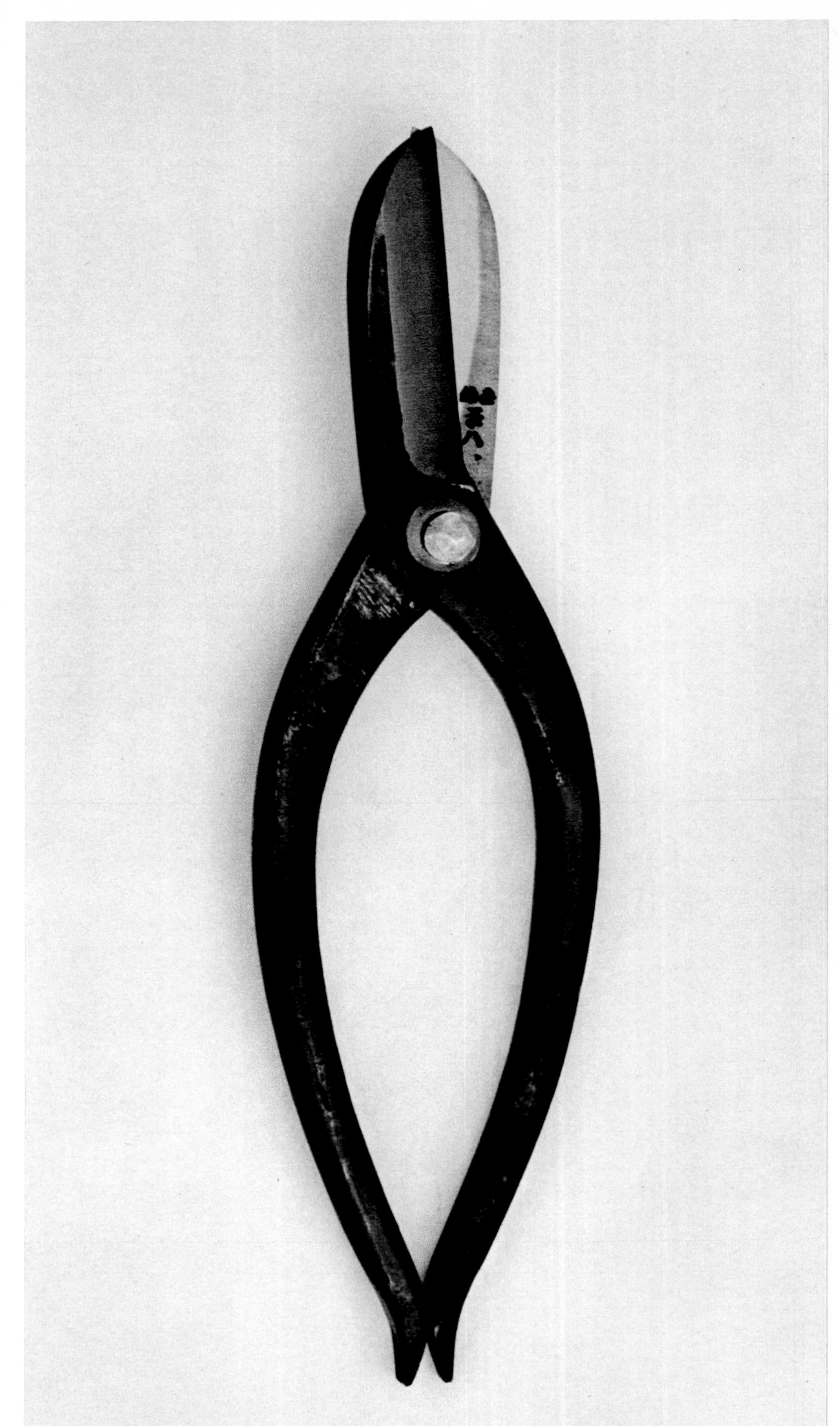

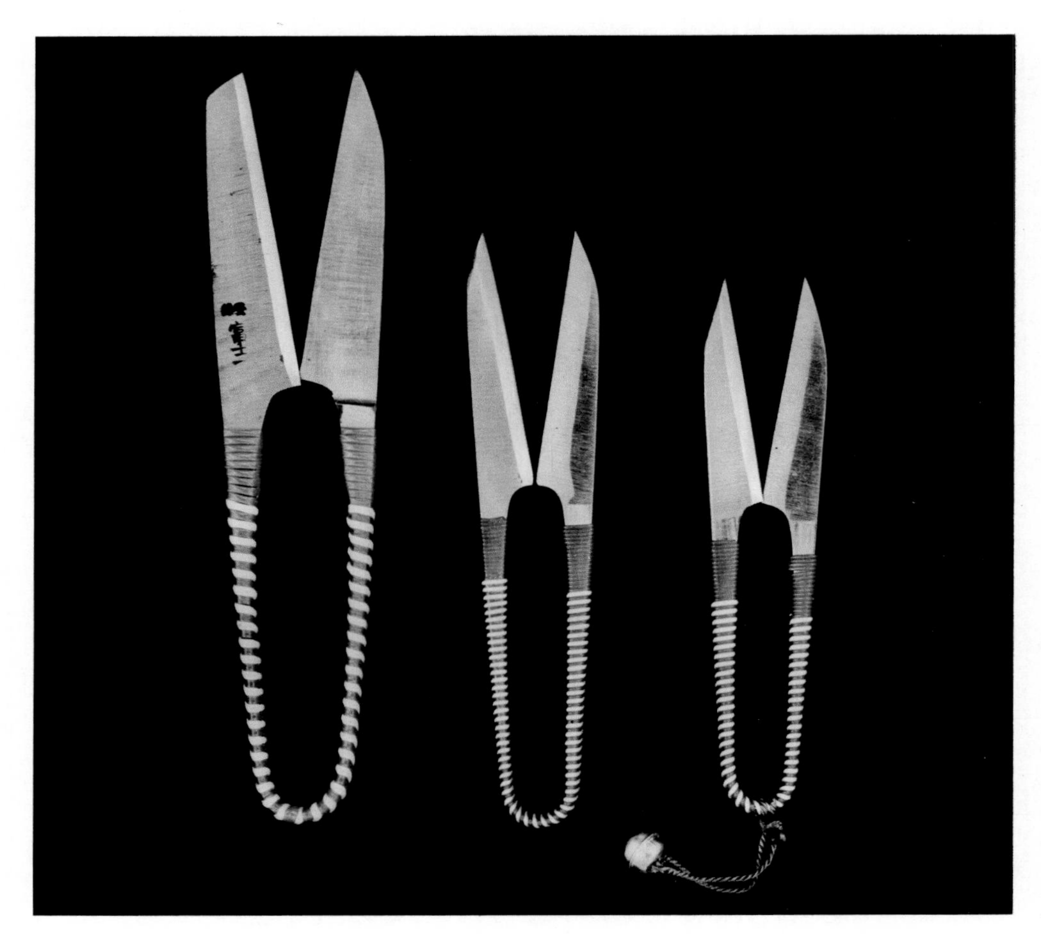

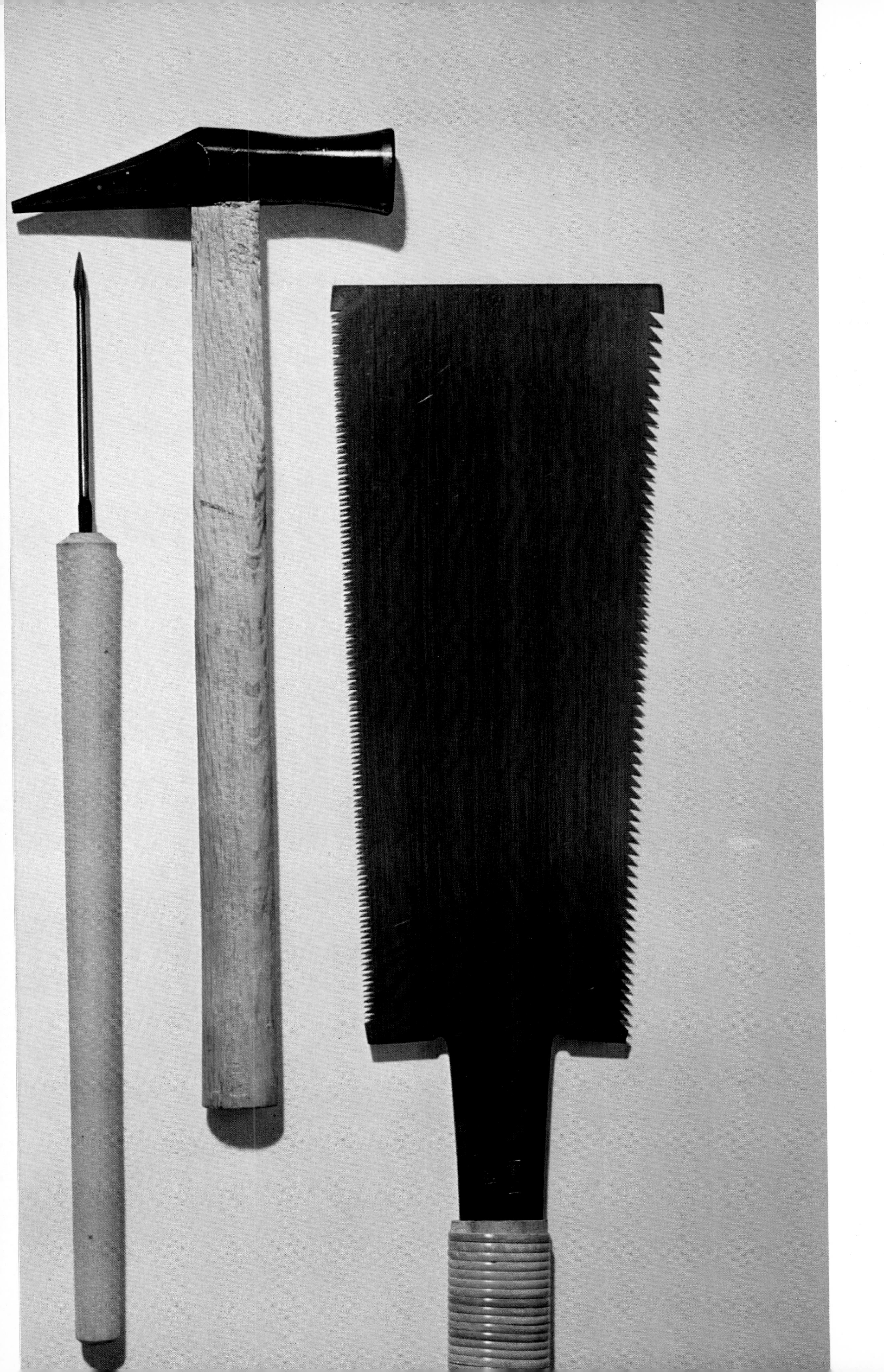

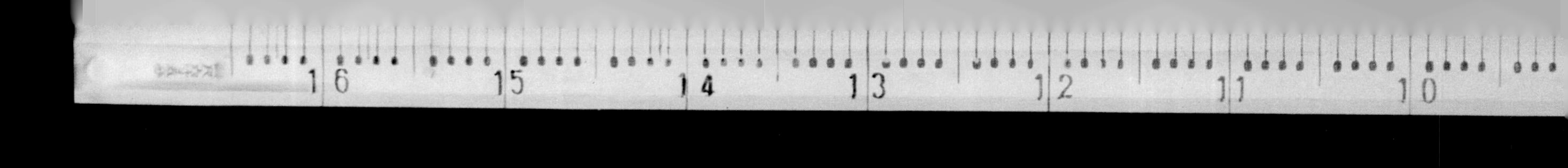
16 15 14 13 12 11 10

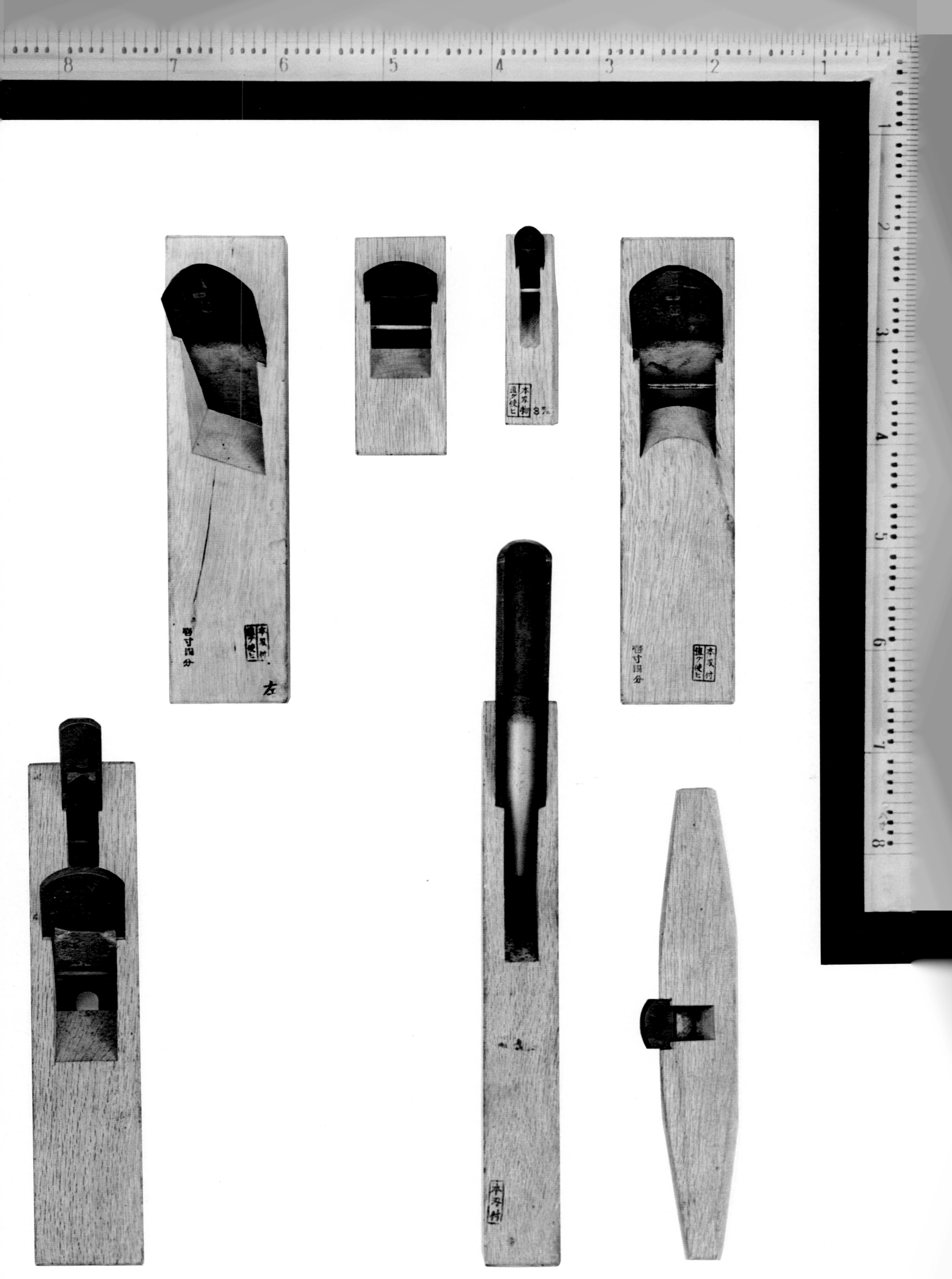

壱寸四分
本刃付
直ヶ使ヒ
左
本刃付
直ヶ使ヒ
壱寸四分
本刃付
直ヶ使ヒ
本刃付

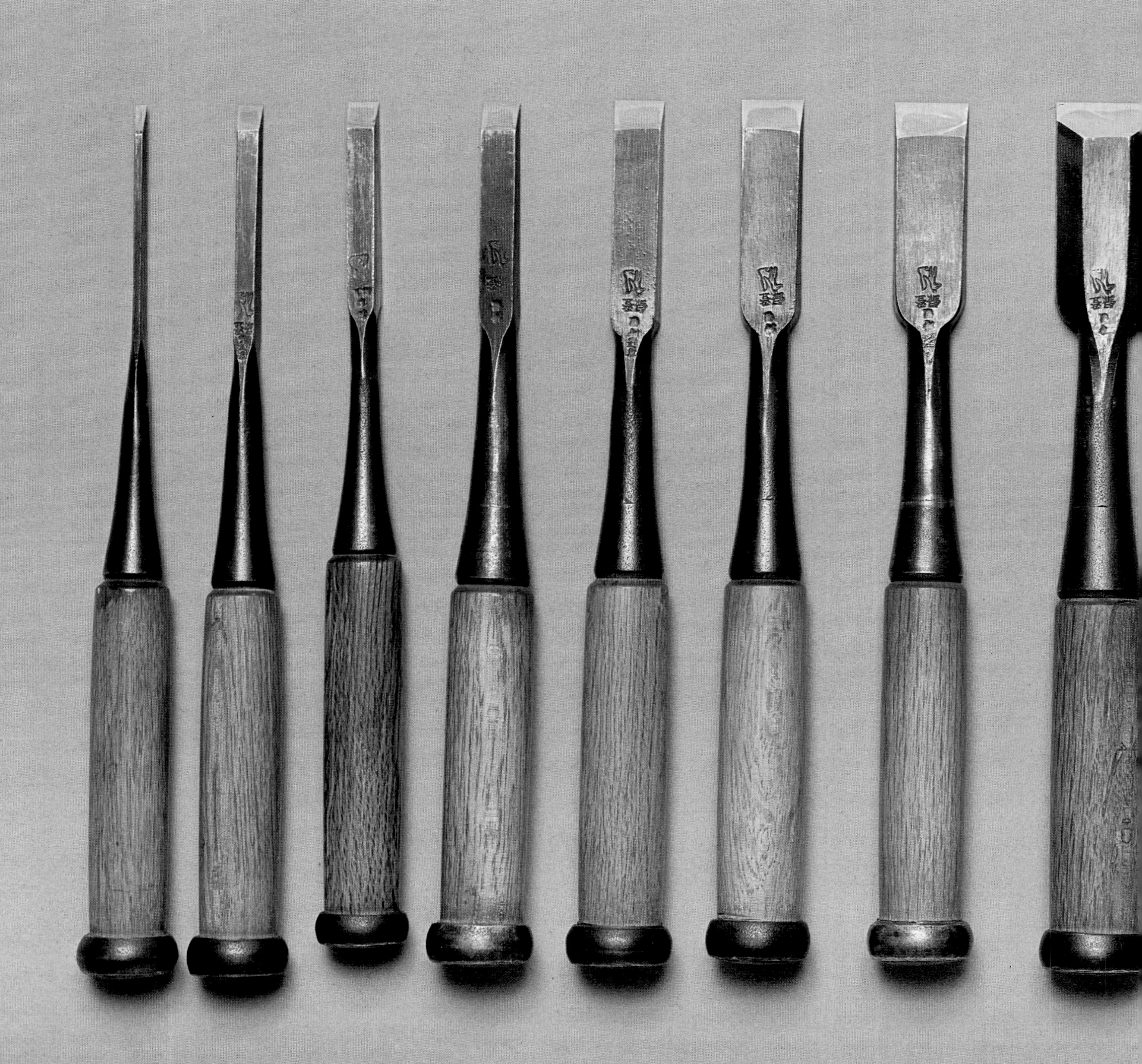

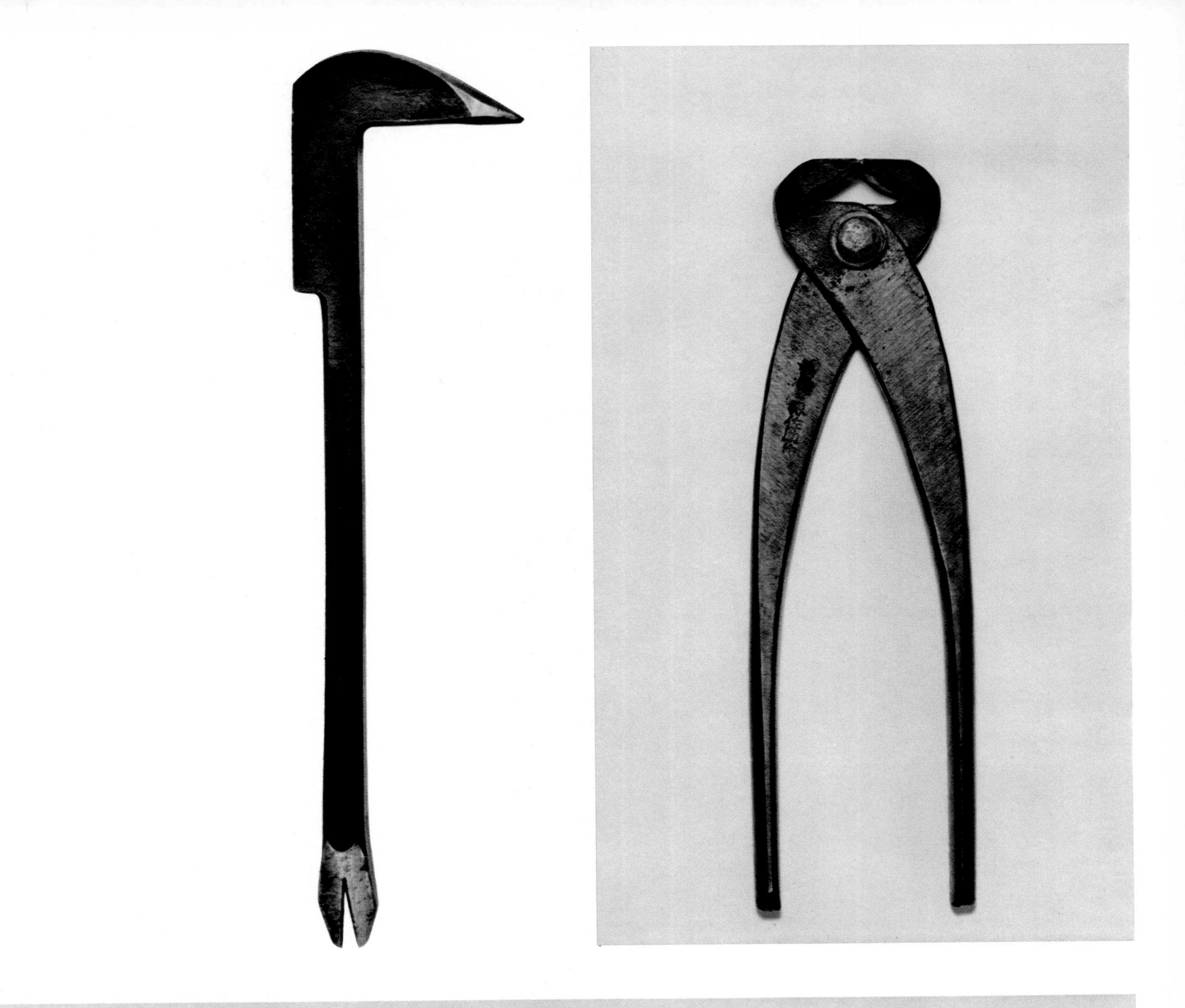

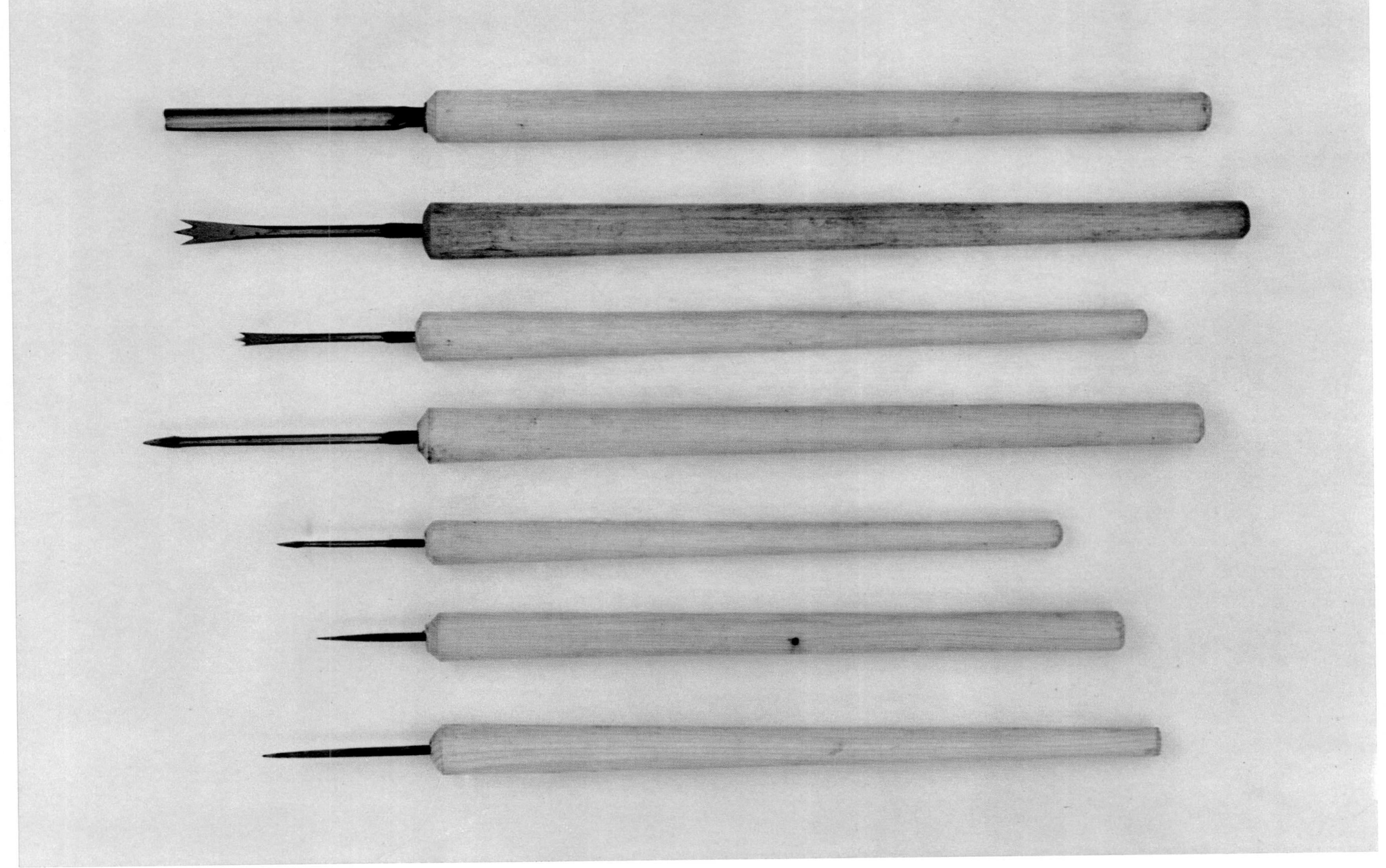

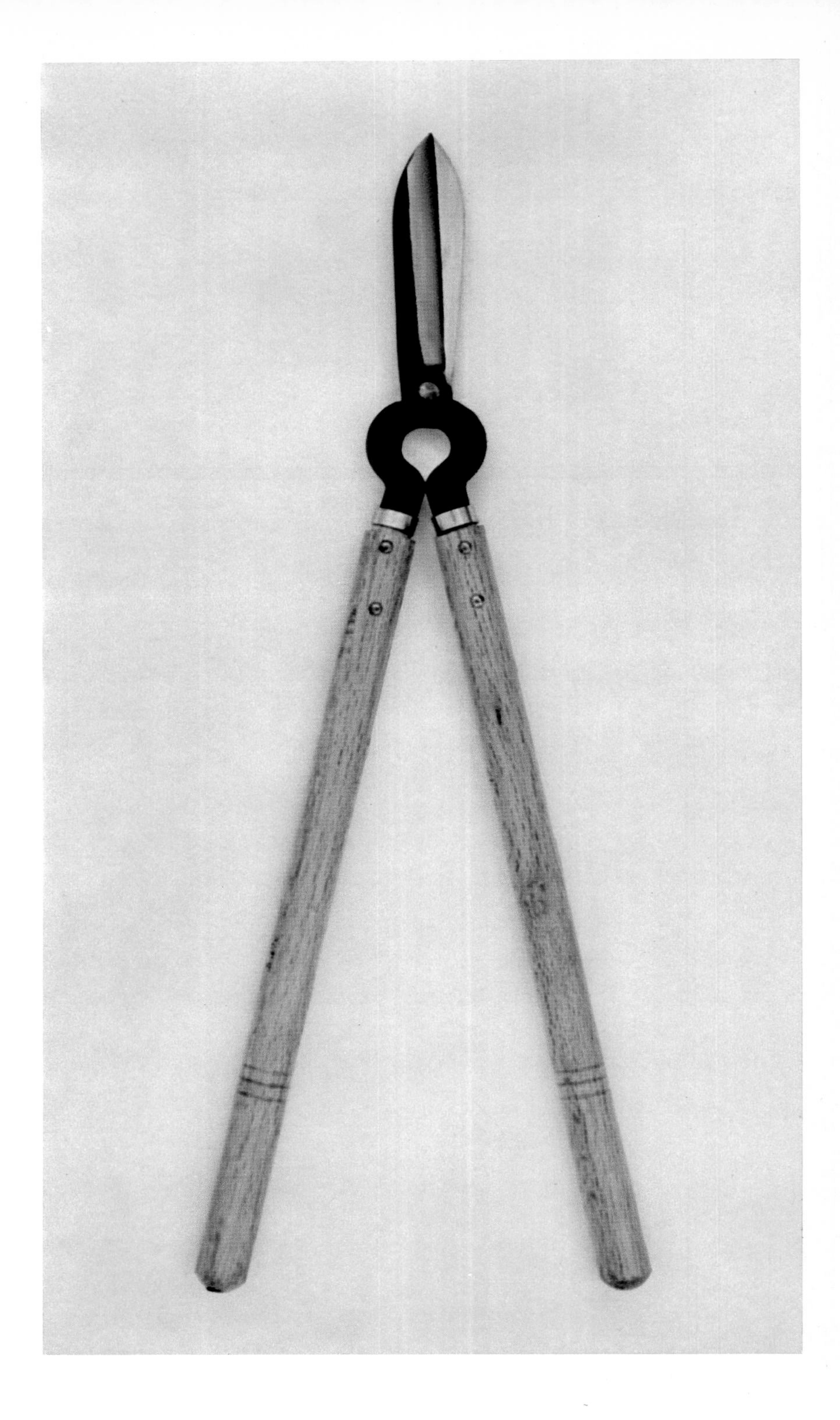

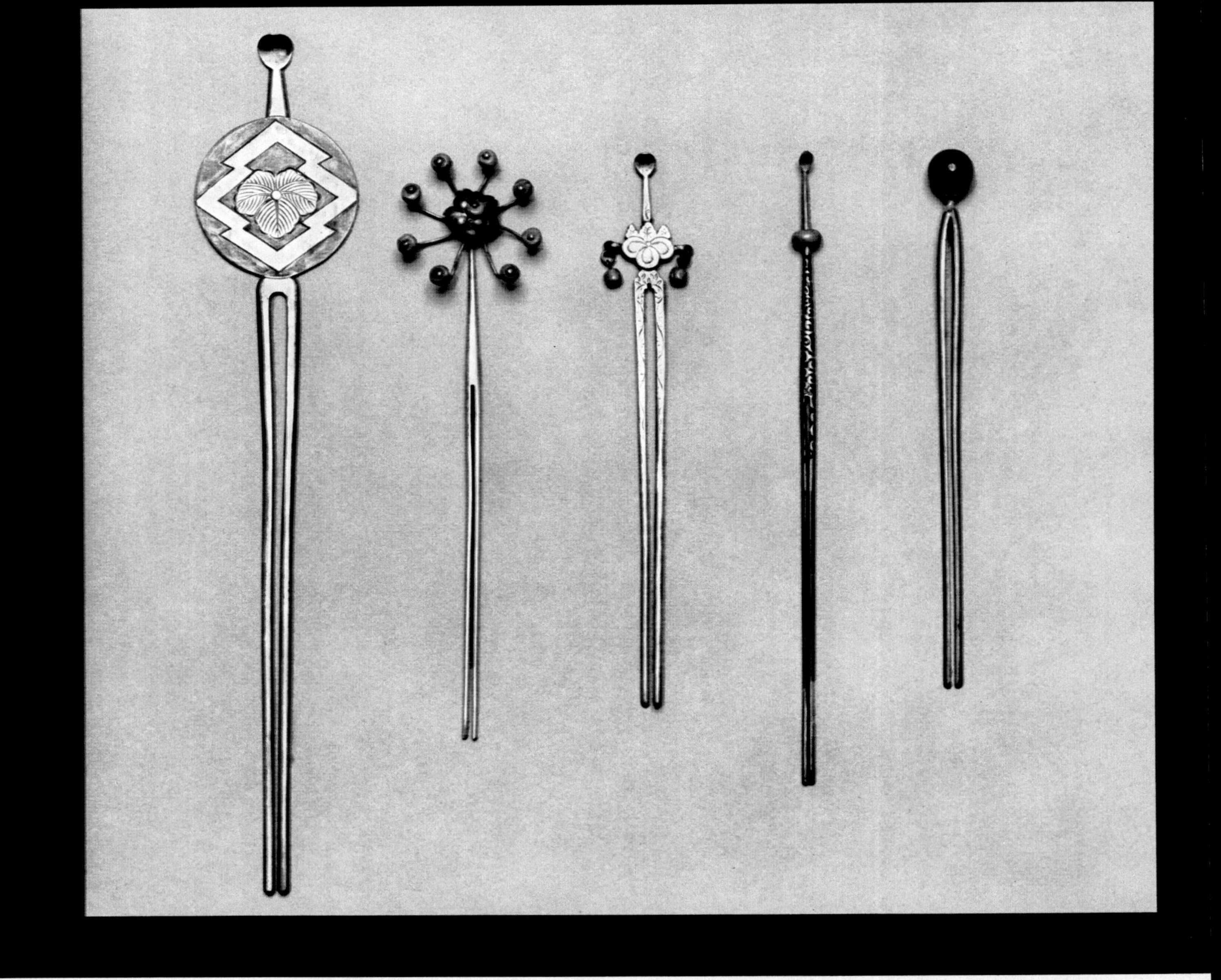

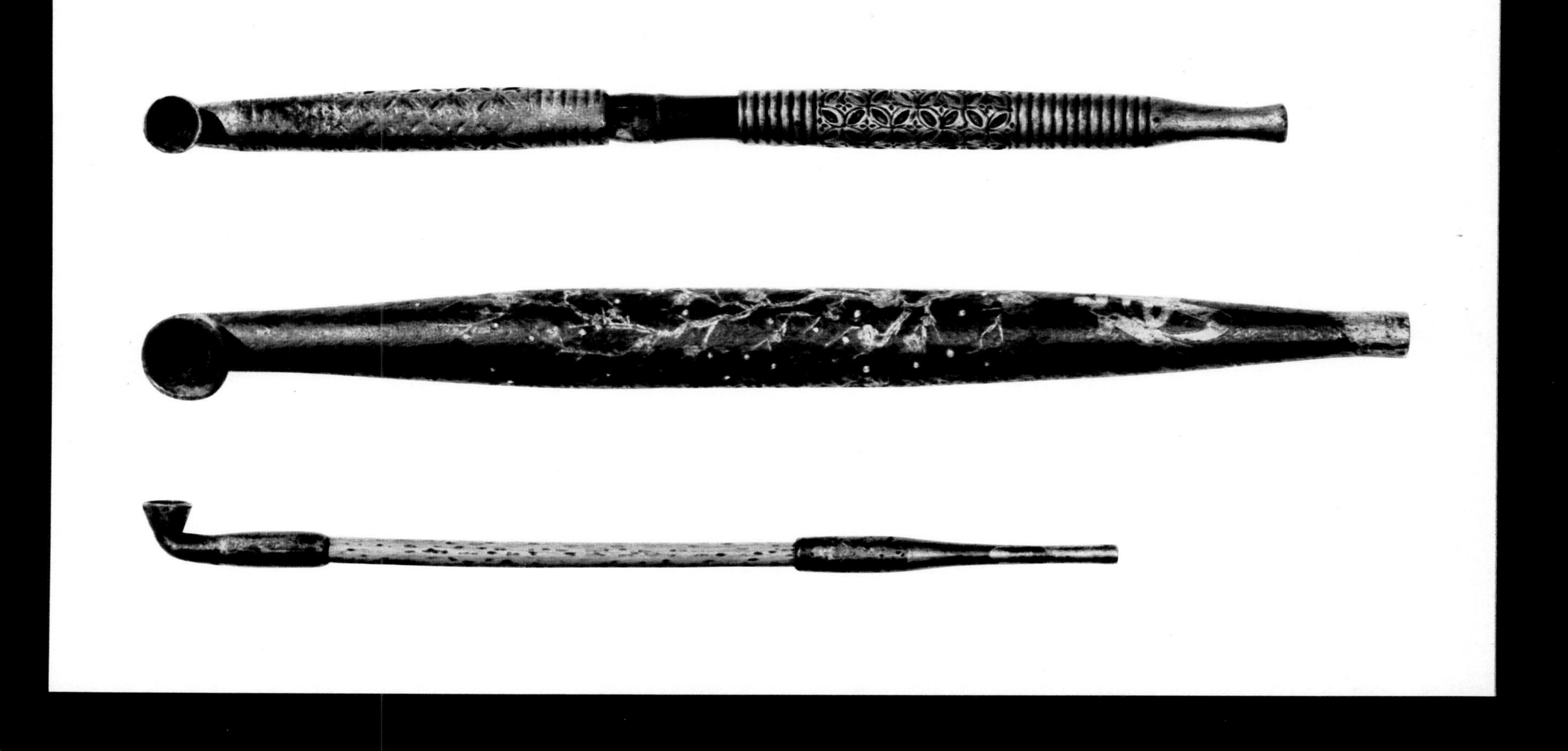

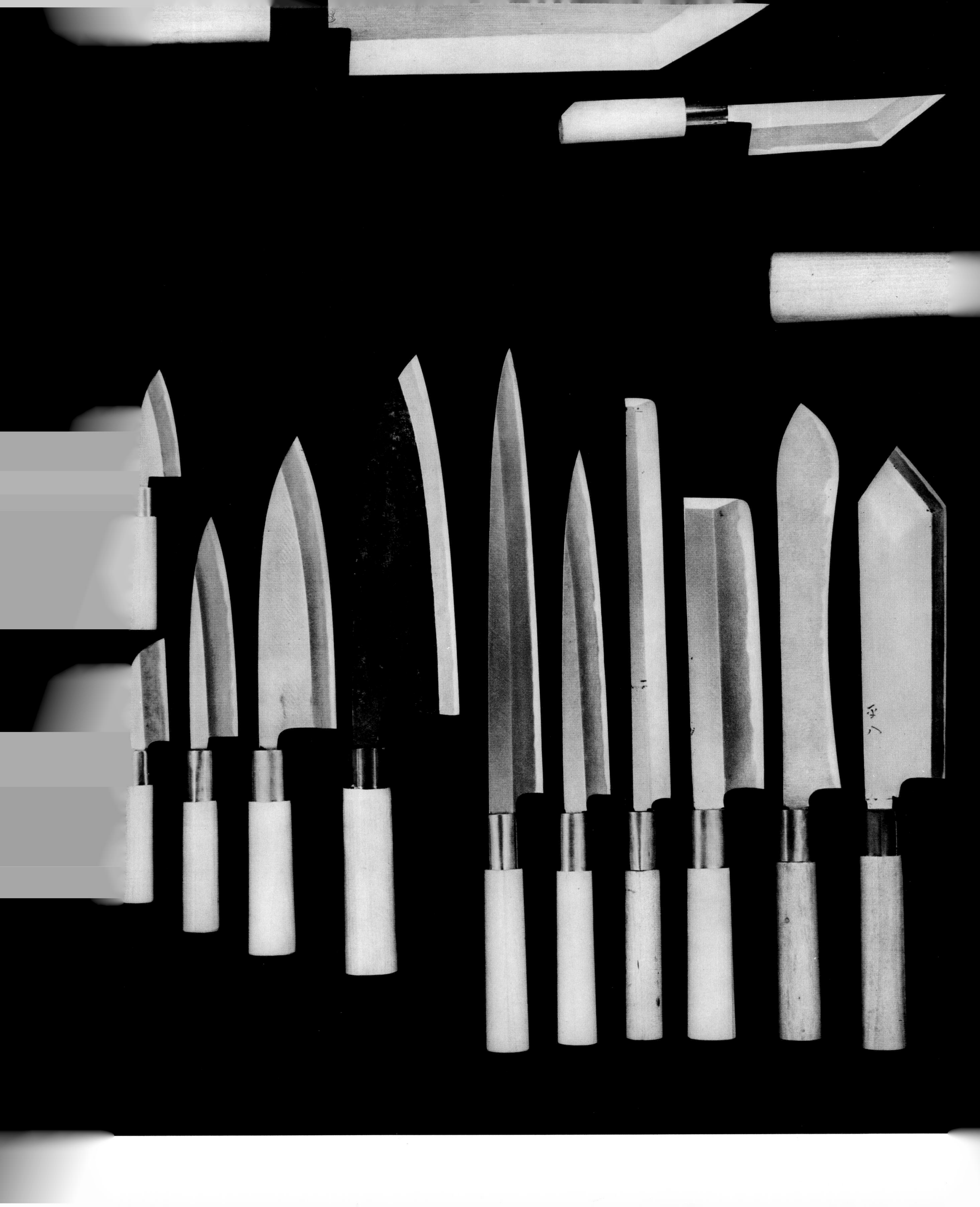

CLAY

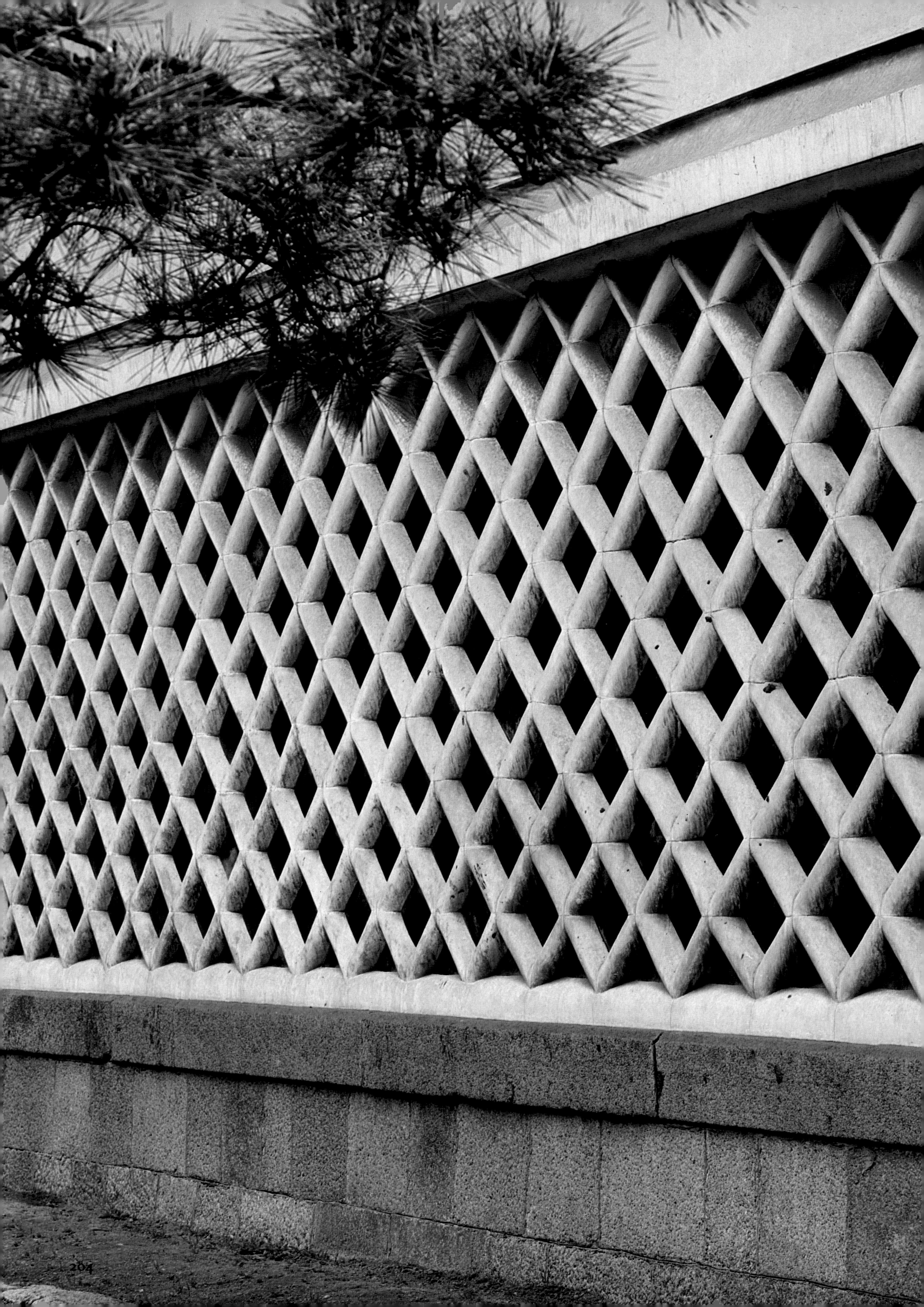

唐招提寺
唐招提寺

唐招提寺

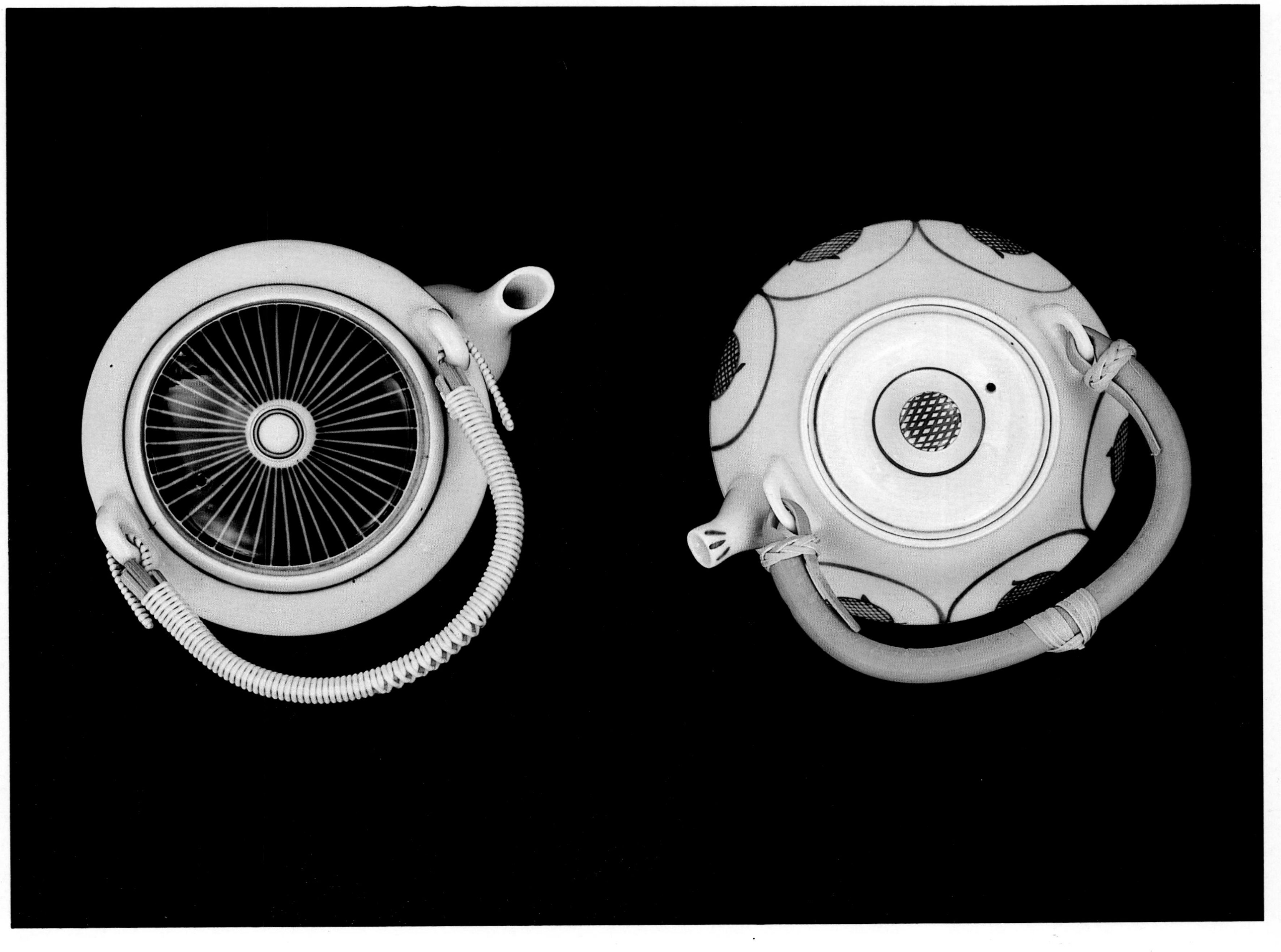

申
未
午
巳
辰
卯

申
未
午
巳
辰
卯

お願
上阪たき江
福井県
南条郡
河野村字糠

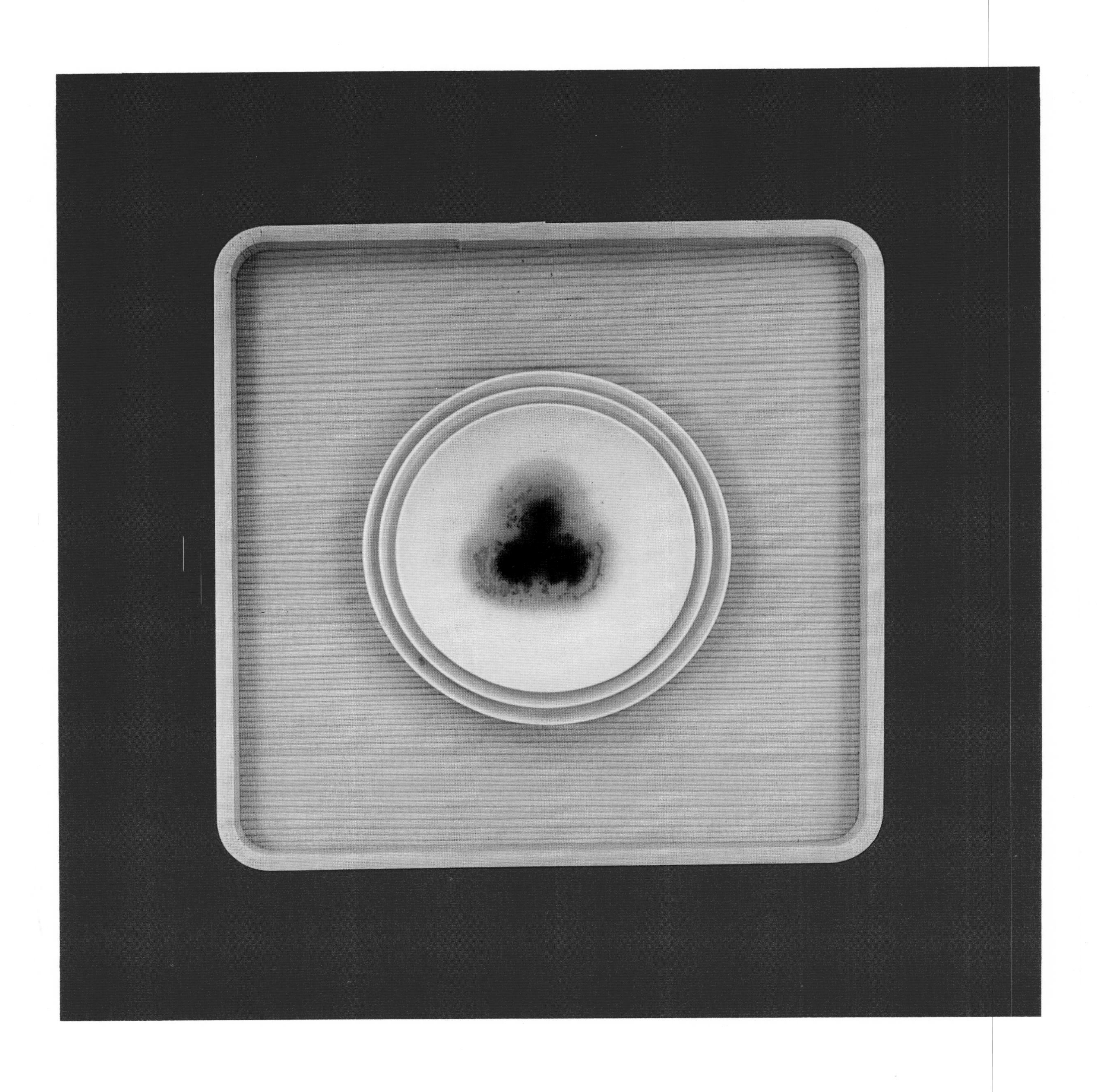

STONE

奉納

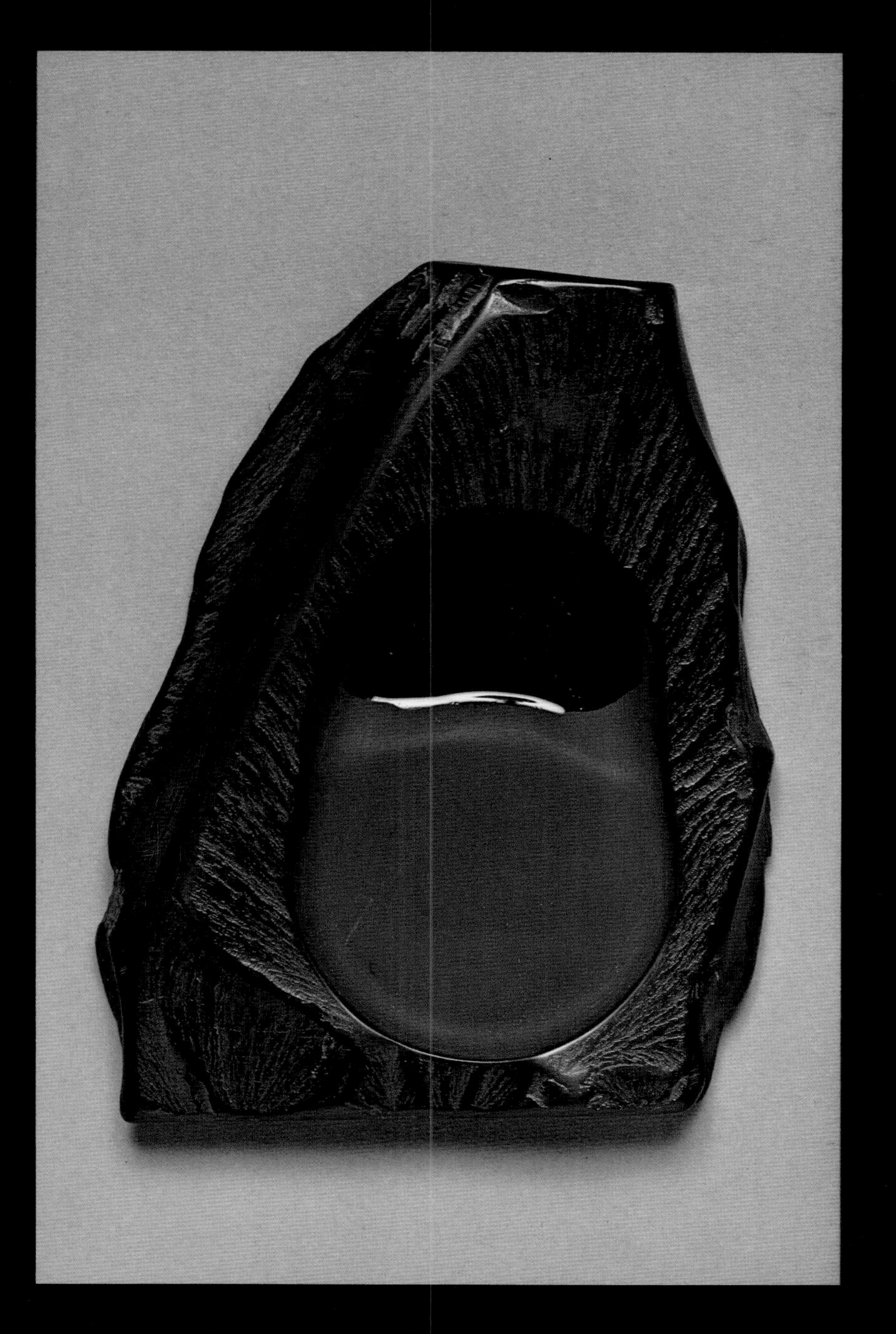

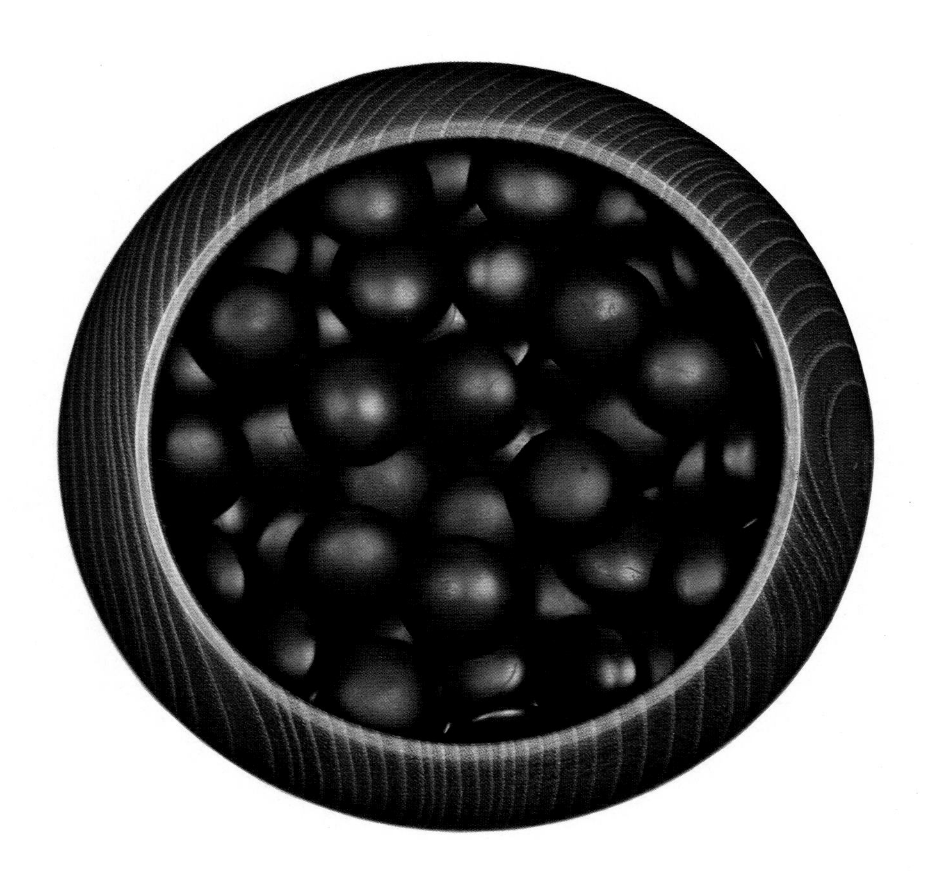

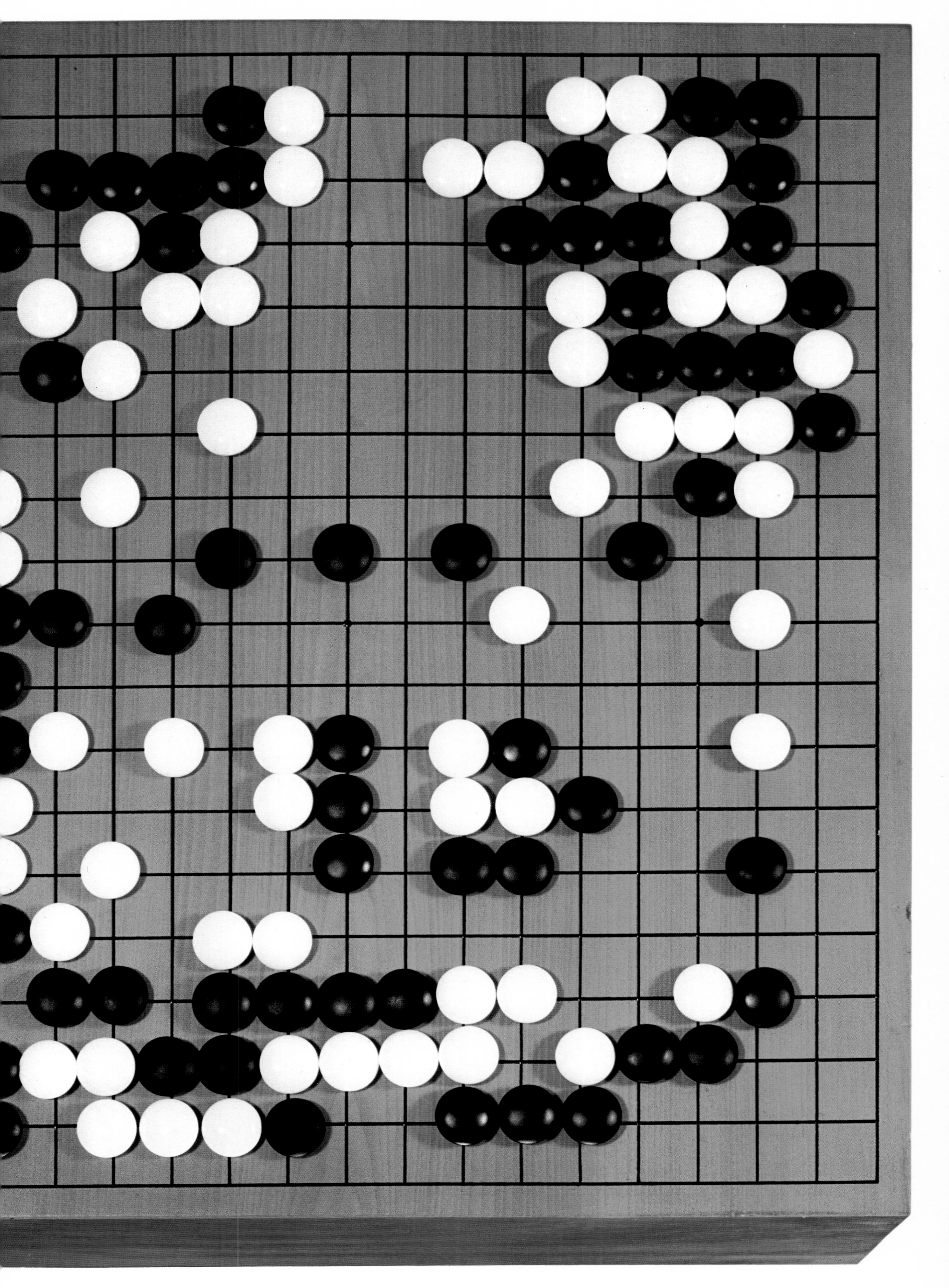

壽

龍

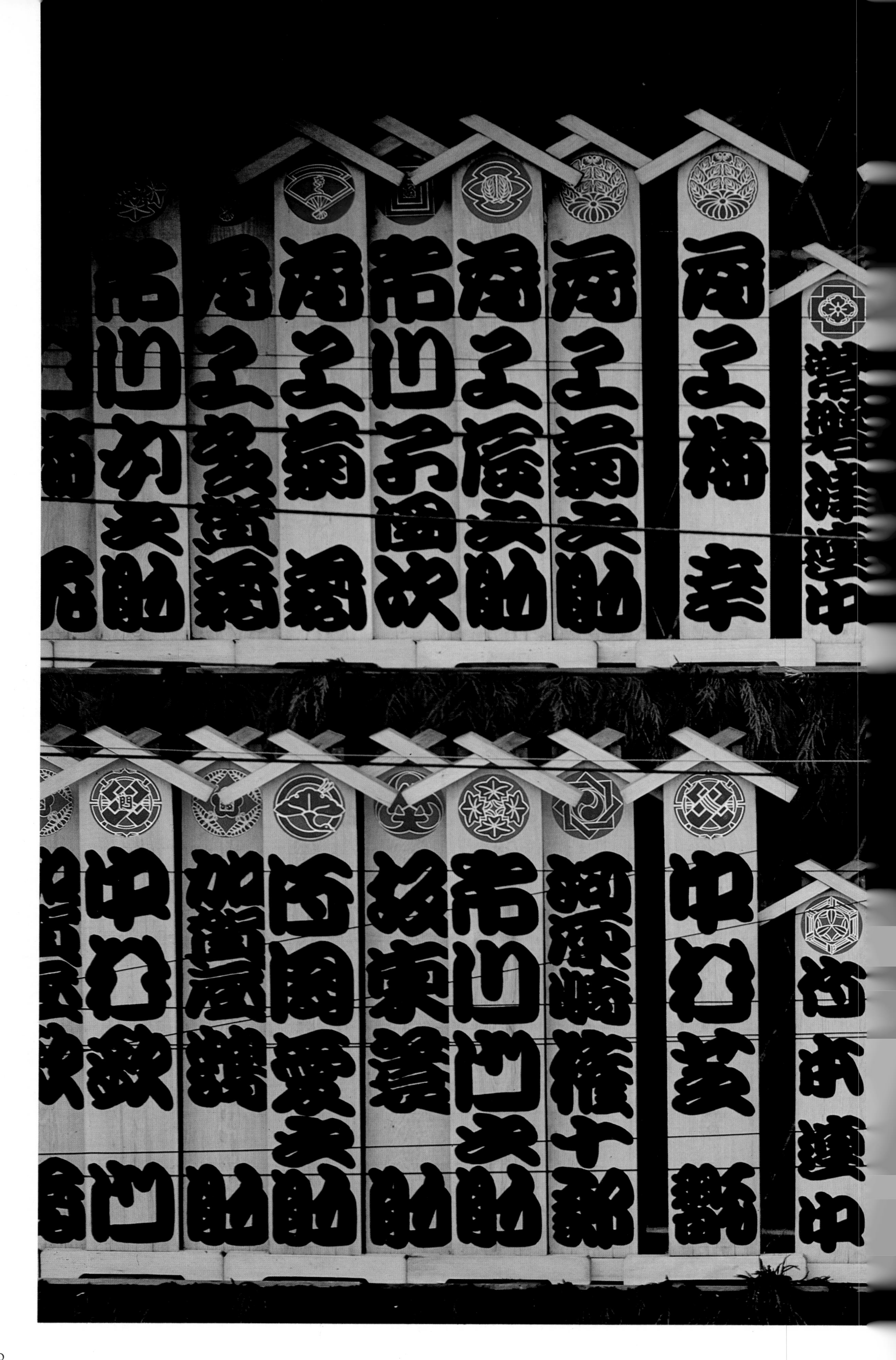
尾上梅幸
尾上菊之助
尾上辰之助
尾上菊五郎
中村芝翫
河原崎権十郎
市川門之助
片岡愛之助

芸術院会員
片岡仁左衛門
片岡秀太郎
坂東三津五郎
嵐雛助
囃子連中

一番

一番

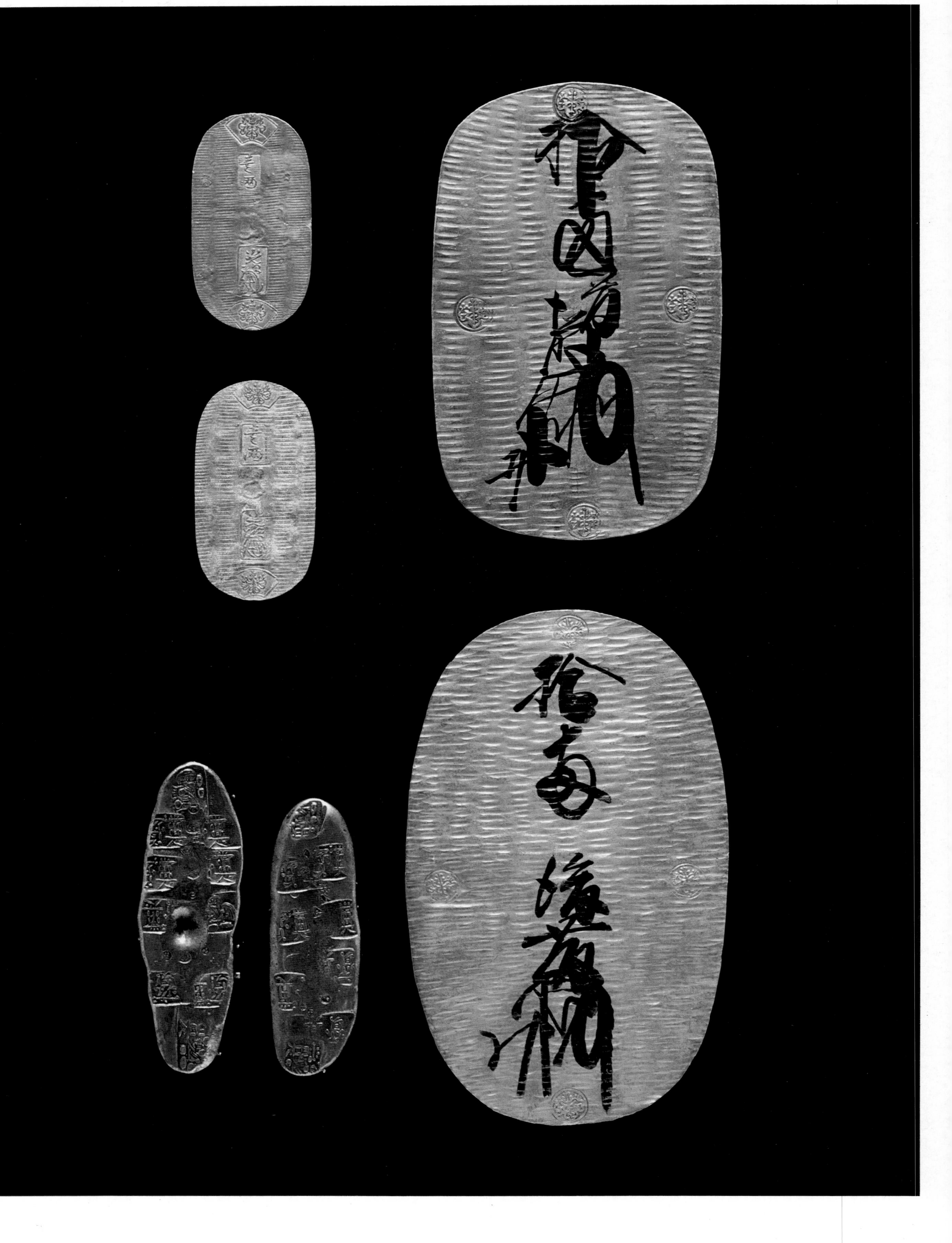

下乘

渡辺
三河三
同朋町
竹本
中川
甘糟
千加坊
元岩瀧
砂場
かじ一
孝吉
つち三
つるた
清一
越武
連和
川正
高石
相撲
茶屋
港町
富士
宇母宗

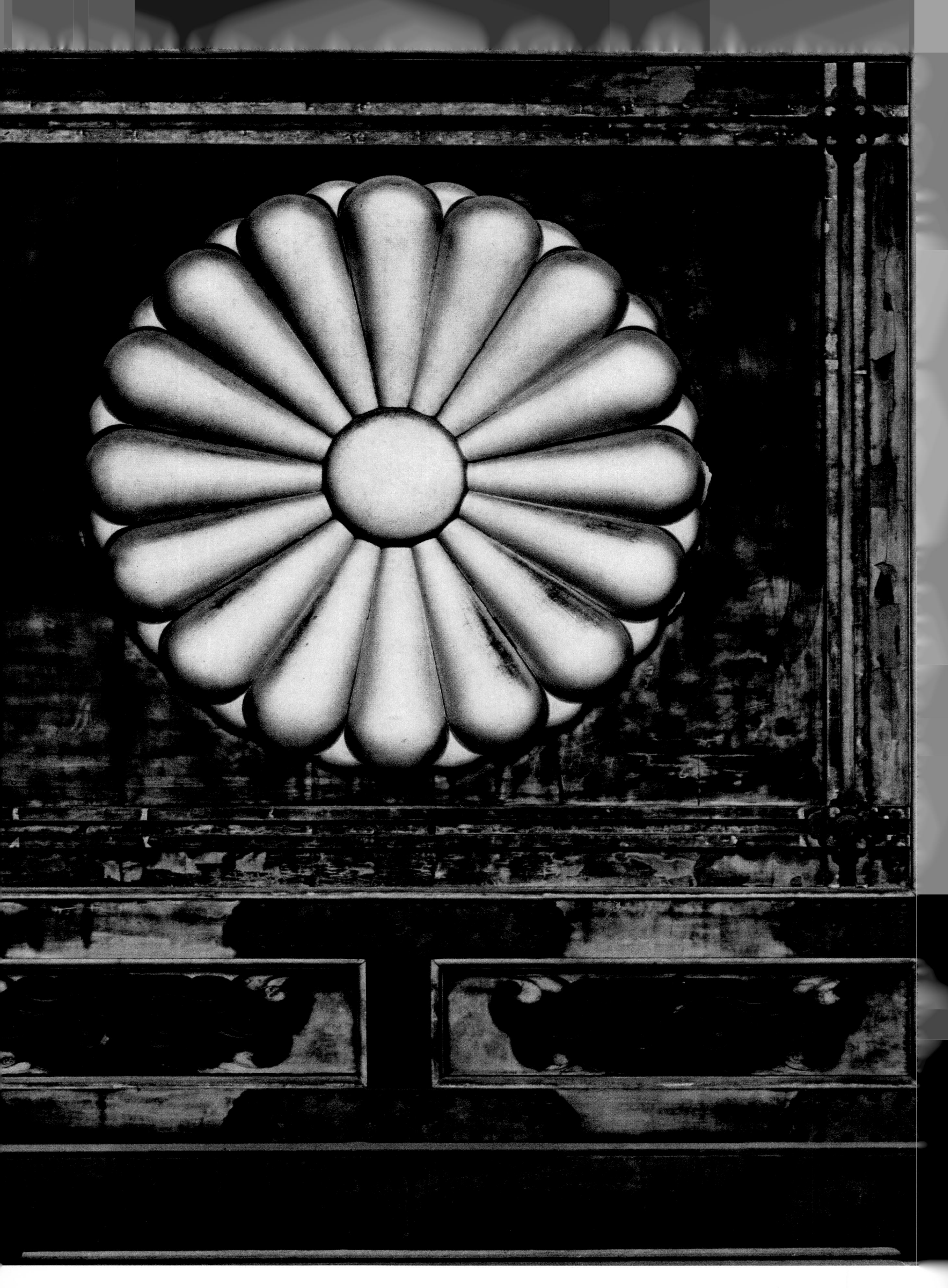

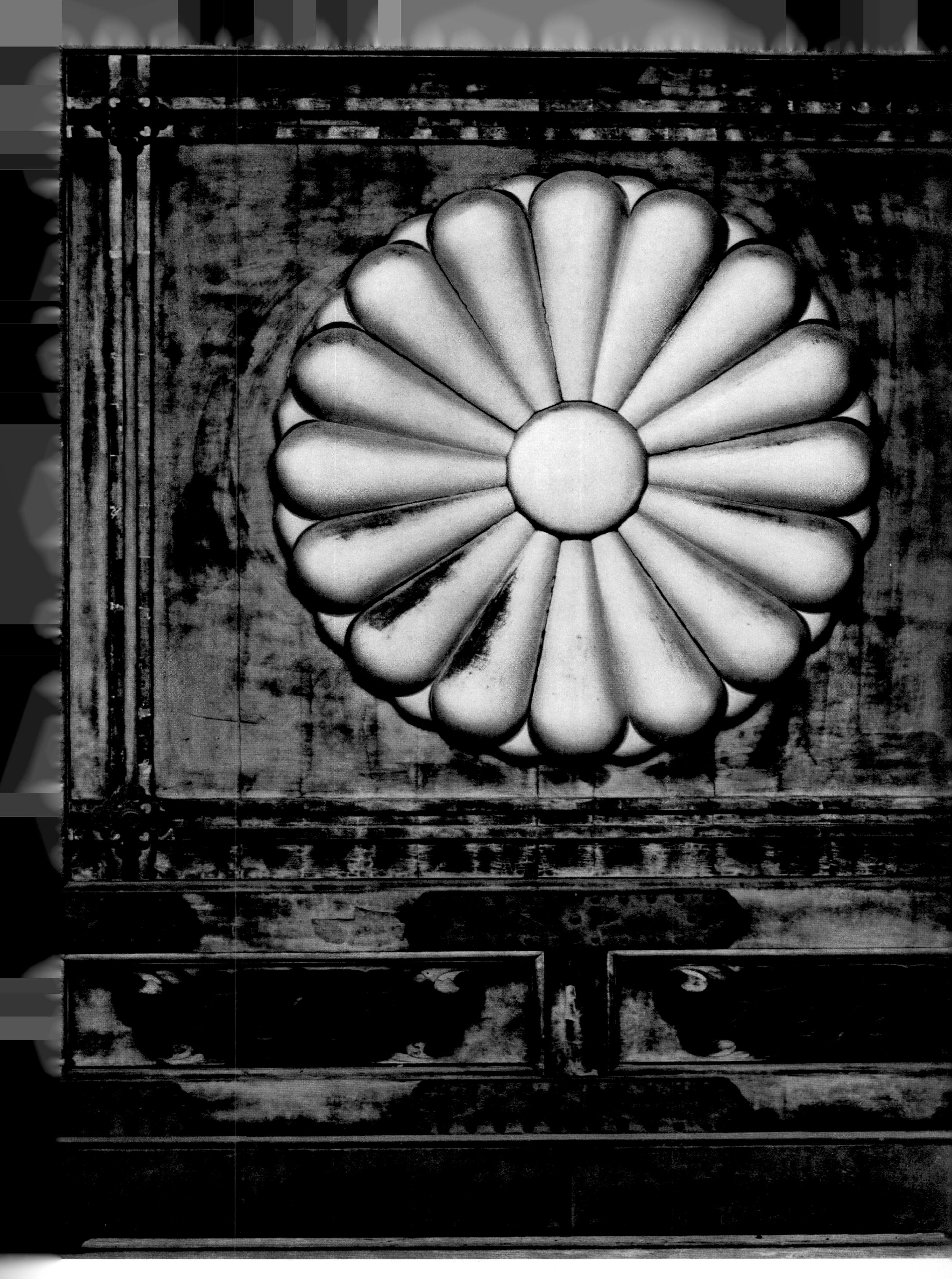

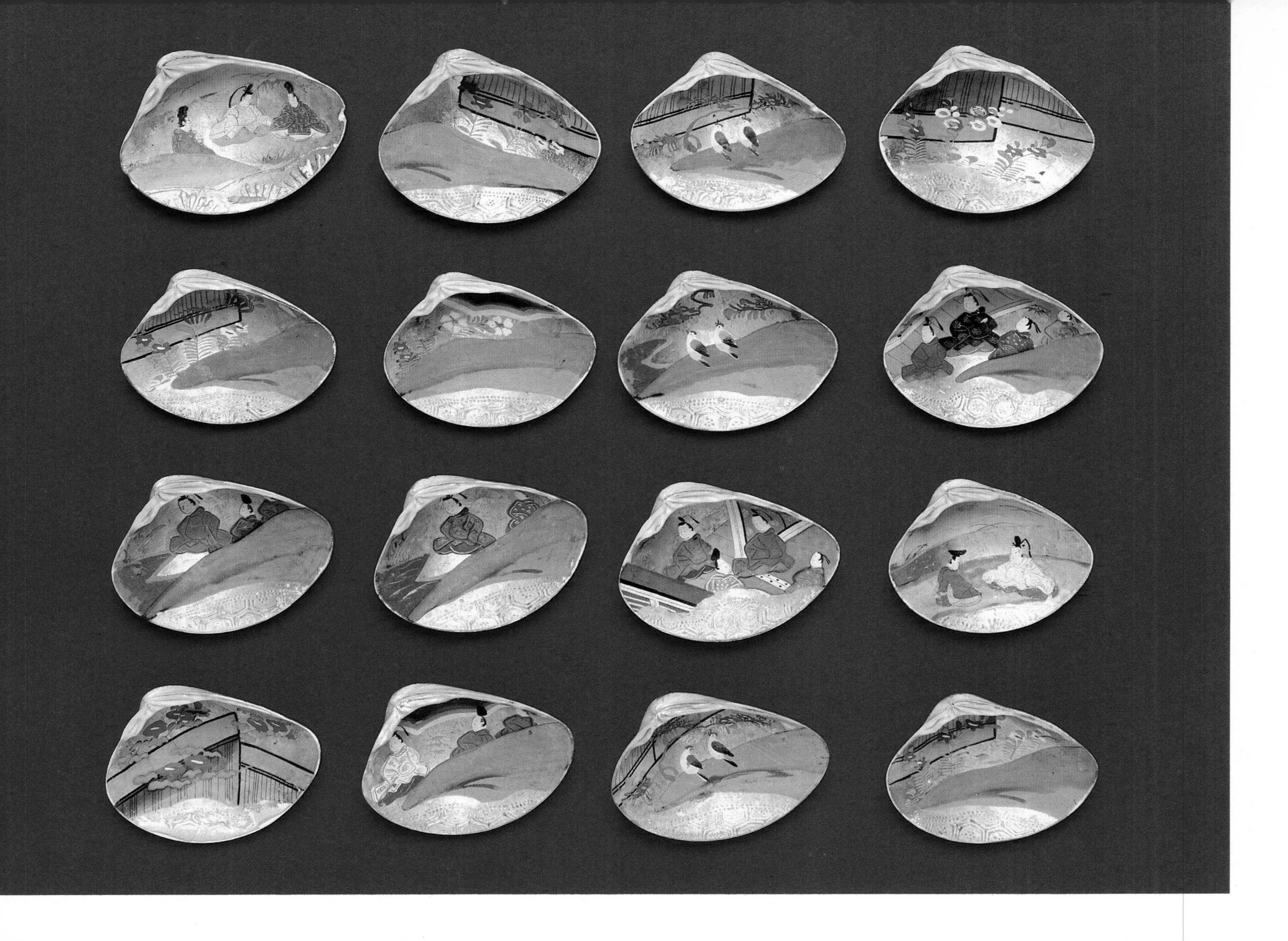

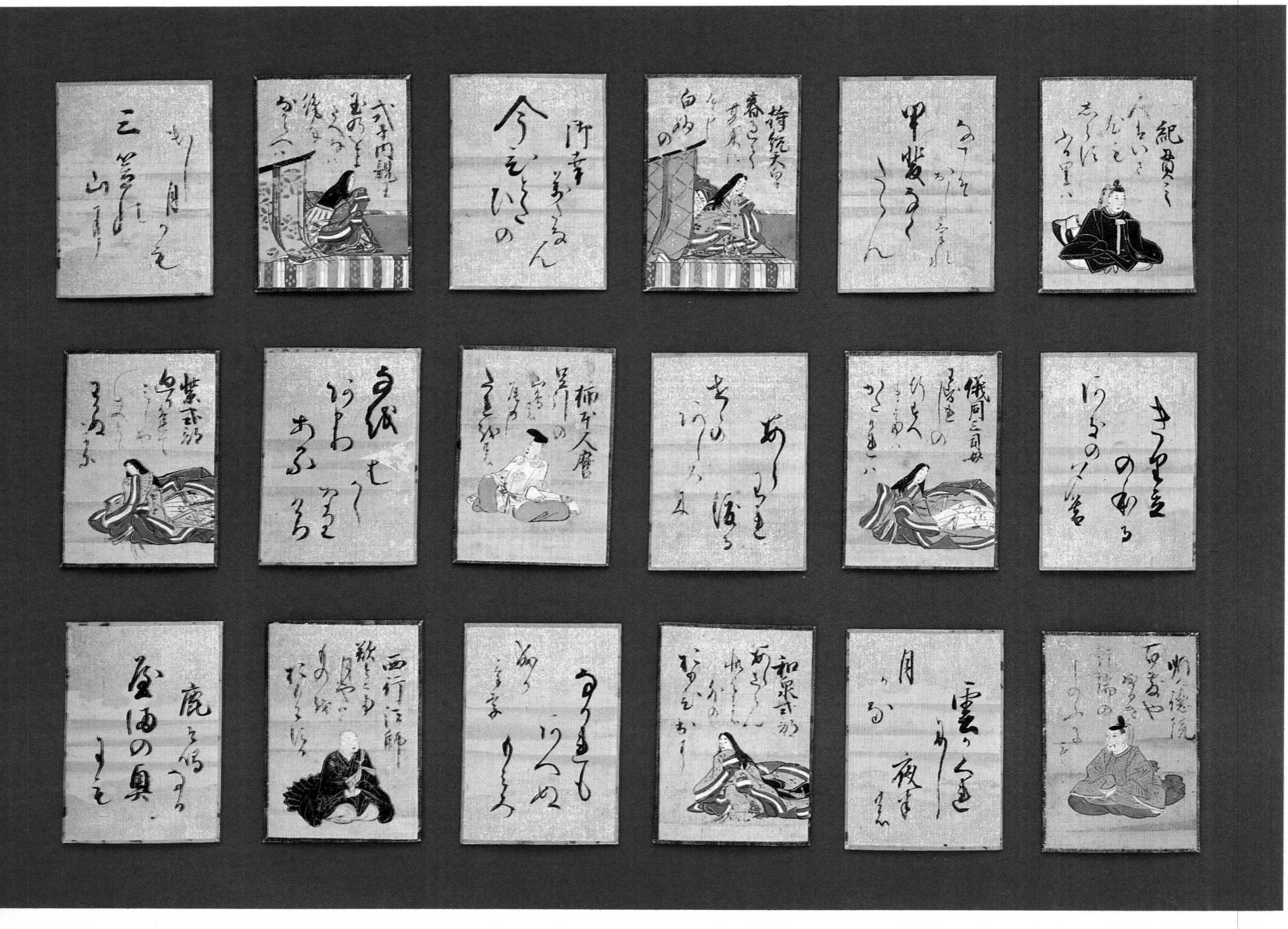
式子内親王
持統天皇
紀貫之
紫式部
柿本人麿
儀同三司母
西行法師
和泉式部

任天堂

The Myriad-Fragment Mirror: An Appreciation

by Richard L. Gage

Given no more than a few unpromising shards of pottery, a piece or two of jewelry, and perhaps some crumbling stonework, archaeologists reconstruct kingdoms that, for want of written record, might otherwise remain obliterated by the vicissitudes of time. If enough artifacts can be reclaimed, the archaeologist's story is a rich one, for peoples are more revealing in the things they make than they are in self-conscious written accounts, often compiled under ulterior motivation.

Japanese society today is neither dead nor moribund. It is vigorously alive and loudly proclaims its own presence and prophesies its own future. But it has changed so much in the past few decades that, year by year, less and less remains of its past. Of course, there is a wealth of documentation to set forth what happened in times gone by, but actual physical remnants of the past are rapidly being swept away in today's tide of change. Fortunately, however, the Japanese people have created a warm record of their past thoughts, acts, and emotions in the objects they have made for use on everyday and holiday occasions. The Japan of times gone is fading fast, and many of the old ways will soon be forgotten. But these many objects constitute a virtual archaeologist's paradise of information with which to reconstruct the Japan of yesterday. This book offers a generous selection of photographs telling part of the eloquent story of the way this people has lived.

For convenience of presentation, the photographs are categorized according to the material from which the objects are made—wood, bamboo, paper, earth, stone, and metal—plus a concluding section on symbols. The remarks offered here, however, attempt to amplify the presentation by discussing the objects from the viewpoints of color, material, and function.

COLOR The Japanese are often regarded by themselves and by others as fanciers of only subtle and refined colors. This is not entirely false. Nor is it entirely true. Anyone who has walked down shop-lined streets in Japanese towns and has been treated to their garish circus of neon, signboards, and plastic cherry blossoms—or maple leaves or wisteria blossoms, depending on the season of the year—must wonder where all the refinement has gone. Or perhaps such a visitor comes to the conclusion that the love of subtlety in color is a thing of the past and that plastic cherry blossoms are no more than a symptom of the universal degenerateness of our times. In fact, both love of the subtle and love of the gaudy are deep-grained in the Japanese aesthetic sense. Whether it is, as some argue, a virtue born of a tradition of repressive governmental sumptuary legislation in earlier times, preference for gray, blue, brown, and muted combinations of such tones rules in mature daily life. Gold brocades, bright crimson, flashy patterns, and other glittering magnificence are reserved for the great and wealthy lords and the imperial court—or at least for their representations on the stage—and for toys and garments of infants or of unmarried young ladies. The tea ceremony, an extraordinary apotheosis of the ordinary, and the deck of cards used in the game called *hanafuda* (flower cards) illustrate these two preferences.

Serving a mixture of powdered green tea and hot water in a calm, refined setting is an ordinary act that, in the tea ceremony, has been elevated to an education entailing a wide range of literary and aesthetic disciplines. Color is one important part of the whole.

The illustration on page 89 gives an idea of the general color setting of a tea ceremony and reveals characteristic preferences for natural colors: subtle, complex colors and limited but effectively introduced accent. The somber black-lacquered board; the black brazier resting on it; the complex tones of the iron kettle; the natural, varied colors of the water container, which is lacquered brown inside; the bamboo of the whisk; the black of the lacquered tea container—all these fit easily into the mood set by the green-tinged beige of the rush matting. A small amount of gold on the tea container gives a highlight that will be further intensified by the tea bowl and the bright green tea in it (page 218).

Examples of fondness for colors and color combinations of this kind are abundant. The natural green of fresh bamboo is much prized alone (chopsticks on page 90) and in combination with aged bamboo (baskets on pages 106 and 107). Admiration for natural colors inspires the Japanese to take a dim view of paint in general, since it conceals the tones and grains of the wood. Their attitude in this connection has the not always happy result of long lines of houses that, blackened by the polluted air in modern cities, look shabbier than they actually are. But this same reluctance to conceal natural tones more often produces effects of great refinement and beauty. No coat of paint could enhance the walls and windows of a tearoom (pages 86–88) or the great variety of traditional bamboo fences (pages 82–85). These bamboo fences provide an excellent example of the Japanese love of color that changes and thus suggests the sadness of the ineluctable passing of things that is a celebrated part of the national psyche. A newly built bamboo fence is entirely, or partly, fresh green (page 84). In a few weeks, or months, it turns to gold (page 85) and then, as it ages further, comes to have the silver-beige look achievable only in the natural course of the passing of time (pages 82–83).

Beige-gray is one of the compound, subtle colors dear to the part of the Japanese mind that is symbolized by the tea ceremony. Other similarly complex hues are found in many different kinds of places. Moss, for example, is beloved for the velvety green it imparts to famous gardens (page 236). But there are many kinds of moss, and one of the kinds most treasured in Japan is the bluish, silvery, green moss on tile roofs (page 97). This moss is often reproduced with great affection on the pine trees decorating the stage walls and sets of the traditional drama. It is hard to define its color, and this may partly account for its appeal. Vagueness and lack of clarity can be observed in many colors that the Japanese favor. Kettles for the tea ceremony, though of iron, are not jet black. They may be greenish (page 168) or red-brownish (page 169). Only cheap ironware is flat black, whereas treasured antique pieces are old enough to have acquired the friendly, indistinct colors that are comfortable to live with.

The kettle on page 168 not only reveals something about complicated colors but also shows what is meant by skillfully placed yet restrained accent: in this case a shiny polished knob on the lid. Knowing where in a somber color composition to put one touch of bright color that brings everything to life requires knowledge, experience, and—most important—flair. Often the Japanese reveal these characteristics to a high degree. The red berries in the flower arrangement on page 96 indicate all three. Vermilion and gold are gaudy, but, used sparingly on the lacquered rim of the hearth in an otherwise somber tearoom (page 167), they do not exist for the sake of their own decorativeness but enliven the total in a way similar to the one in which the red lips of the Noh mask (page 57) and the pinkish accents of the Bunraku puppet head (page 60) suggest living flesh in dead wood and paint.

When they are determined to be gaudy, however, the Japanese are peerless. The brilliance of the effects they achieve arises from a number of technical approaches to the employment of color and pattern. The flower cards, or *hanafuda,* on page 279 illustrate three such techniques: bold areas of flat, pure color (the white moon, red sky, and black pampas field in the fifth card from the left in the second row); elaborate, multicolored patterns; and color used symbolically (for instance, the little yellow tips in the fifth card from the left in the first row identify the black objects as pine boughs and not, perhaps, squirrel tails). Bold areas of pure color produce dazzling, sometimes incandescent, effects, especially in the settings and costumes for the Bunraku puppet drama and the Kabuki. For example, the red in the kimono of the puppet at right on page 61 is more intensely alive because of its proximity to the blue in the sleeves. When costumes of this kind are seen under the bright lights of the modern stage, a kind of visual blurring occurs to make boundary lines among colors indistinct. No doubt in the past, when theaters were less brightly lighted than they are today, manipulation of color both enlivened the scene and symbolically enlarged the persons of the performers. Umbrellas (pages 110–11 and 137) illustrate the vivid use of large, bold areas of color, as do many kinds of lacquerware. In lacquered dishes the most frequently encountered colors are black and red (page 25), both of which have special festival significance. Other colors, however, are sometimes used, as shown in the tiered box on page 50, which suggests the striped pattern of the curtain traditionally used on the Kabuki stage.

Elaborate or even fussy patterns, usually combining natural forms, or stylized versions of them, and many colors, abound in Japanese art and daily life. Such elaboration can be seen in the fans on page 138, the clothing of the dolls on page 144, and notably the many-layered costume of the Heian period (794–1185), represented by the garments of the Girls' Day *hina* dolls on pages 66 and 67. Though splendor of this kind, as well as the vermilion, green, and blue in which Buddhist temples were originally decorated, may have been partly the result of Chinese influence, the taste for them took root in Japan and persists today in decorations for the grand, the festive, and the juvenile.

Just as gaudy magnificence and richness symbolize wealth and power or the fleeting happiness of childhood, other colors are frequently charged with meaning. Red and black, as has been noted, are festive colors employed in the lacquerware from which holiday food is eaten. Interestingly enough, black is sometimes also a funereal color in Japan. Again, on the Kabuki stage, young women often wear black brightened with red and purple. Geisha favor the black kimono with decorative patterns for New Year's, and a woman usually has in her wardrobe a black kimono with an embroidered or otherwise ornamented skirt to be worn with a rich brocade obi to weddings.

White is a color of purity, and, given the traditional Japanese reverence for cleanliness, this means a color of sacredness. The folded-paper pendants (page 160) that signify sanctity—for instance, the sanctity of a Shinto shrine compound—are usually white, as are the utensils in which offerings to the Shinto gods are made (pages 230–31) and the small cups from which ceremonial sakè is drunk at Shinto-style weddings (page 232). Unfinished wood, much prized in Japan and employed in all kinds of holy articles and architecture, is called *shiraki*—literally, white wood.

On page 179 is a lock mounted on a set of fancy woodwork doors at the famous Tokugawa mausoleums in Nikko, where the buildings are so covered with carving, gilding, and a general clutter of decoration in a riot of colors that architecture itself virtually vanishes. Decades ago it was fashionable to ridicule the Nikko shrine buildings as vulgar display and to hold up as their opposite, and thus as the paragon of good taste, the refined simplicity—fussy, too, in its way—of the Katsura Detached Palace, in Kyoto. The two places represent the two strands of the Japanese interpretation of color, and it is noteworthy that Japanese tourists continue to dote on both.

MATERIALS Two aspects of a material help to determine the forms of the objects that are made from it: intrinsic physical traits and the nonmaterial characteristics imposed on the material by the interpretive acts of man. For instance, a hard, roundish stone, when tied to a stick, is a hammer for obvious reasons. Little imagination is required to see a scoop or a bowl in a bivalve shell of convenient size. A large leaf unmistakably offers itself as a dish or a wrapper. Probably wherever he has gone, man has put objects to the same or very similar primitive uses. Nonmaterial characteristics, on the other hand, vary much more, depending on the way man has reacted to his own distinctive historical experience. Because of its relative rarity, gold has long been one of the bases of human commerce in many parts of the world, but the Incas of Peru are said to have valued it only for its ornamental worth and to have been perplexed, possibly amused, by the Spaniards' great greed for the pretty yellow metal.

In some cases, however, even the purely physical nature of the material suggests function only to a sensitive and practiced eye. In other words, the trained craftsman applies his insight and his tools, which through long familiarity have become virtually extensions of his own limbs, in the act of liberating form from the natural material that figuratively encases it. Various peoples combine these three approaches to form in various proportions that can symbolize a general approach to the world and the place of mankind in it. Some emphasize the element of human interpretation to the point where the materials are put to use in forms that have little in common with their natural conditions. The Egyptians made stone lotus buds for the capitals of their columns, and the Greeks translated an architectural system that made good use of the natural traits of wood into marble, a material with which it made much less sense. The Japanese consider themselves more one with nature than nature's master. It is not surprising, then, that in Japan, in the majority of instances, the element of human manipulation, though ubiquitously present, has been subordinated to the character of the material. For this reason, the Japanese people have been extremely sensitive in discovering forms consonant with the natural qualities of the materials that they have found most compatible.

The physical appearance of metal is perforce drastically altered in processes whereby the material must be worked, but it has certain physical traits— malleability, castability, and colors—that the Japanese have put to excellent effect. The Japanese sword (page 176) and the guards used with it (page 177) reveal a high caliber of skill and understanding. Experts consider the blades of Japanese swords to be among the finest ever made. A firm grasp of the strength and temperability of metal distinguishes the wide range of excellent tools employed by traditional craftsmen (pages 182–93). Sophisticated metal-casting techniques that produced colossal Buddhist statues and immense temple bells in the past continue to be appreciated today, especially in the manufacture of such things as braziers and kettles for the tea ceremony (pages 167–71). Kettles of this kind illustrate the Japanese fondness for the subtle variation of color producible in cast iron and bronze. In general, glittering metal is not much loved by the Japanese, although hair ornaments (page 194) and the wildly fanciful decorations of some helmets (pages 174–75) reveal the love of richness and display that has already been discussed.

Clay must be so processed in the production of ceramics that little of its original form can be said to persist in the finished ware. Its colors and textures, however, can be, and in Japan are, appreciated and employed to excellent advantage. Many everyday dishes are glazed and decorated in ways that obscure the basic clay in the same manner that paint obliterates wood (pages 210; 211, bottom; 212, bottom). But in other vessels (pages 211, top; 214; 223) the color and texture of the clay set the mood of the entire piece. A happy medium between these two approaches is found in many of the bowls used for the tea ceremony (pages 218–19). In these vessels, decorative glazes and patterns partly conceal the clay, which nonetheless makes itself readily apparent in the form, streaks, grains, and cracks that are knowingly admired by connoisseurs.

Hardness and recalcitrance to processing may account for the low popularity of stone in most aspects of Japanese daily life, although important exceptions are the stones on which ink sticks are ground (pages 250–51) and the polished stones used in the game of go (pages 254–55). Except for some rare cases (page 256), stone sculpture was practically unknown in the past. It is true that in fairly late times grist mills (pages 252–53) found a place in the ordinary domestic establishment, but other articles of processed stone remain rare to this day.

Durability and hardness, the very physical characteristics that have made stone difficult to work and have therefore limited its use for everyday objects, make it eminently suited to such things as castle ramparts (pages 246–47), garden walks and steppingstones (pages 234–36), and staircases (page 237). Its semipermanent nature has given stone its firmest place in the Japanese system of symbols. The obvious venerability of massive old boulders inspired the Japanese of the distant past to include them in the highly elastic and inclusive Shinto pantheon. Like trees and other natural phenomena, old stones are frequently seen bearing ropes and paper pendants indicating their holiness. The sanctity of the material carried over into the creation of stone garden groupings representing three Buddhas or other revered or holy beings. Because it seems eternal to the human mind, stone is the perfect material for memorials to the dead (pages 242–43) and for basins for water to purify the mouths and hands of people about to take part in a tea ceremony or to worship at a temple or shrine (page 233). Finally, since it is practically impervious to destruction by fire, stone serves for lanterns holding the flames that light temple yards and gardens (pages 238 and 240), and, relatively immune to the ravages of time, it is the proper material in which to carve the footprints of the Buddha (page 241).

If stone is the least popular material among the Japanese for objects of everyday use, wood is undeniably the most popular. It has given them their homes, their footwear, their dinnerware, their masks, their combs, their musical instruments, and an almost endless list of things useful, beautiful, or otherwise valuable.

Sometimes wood is visually obliterated by many coats of shining lacquer in red, black, gold, or other colors. But even then its presence is proclaimed by lightness of weight or other physical qualities. Often, however, the Japanese prefer wood either unfinished or finished in such a way as to reveal the much admired grain. Unfinished, or "white," wood, associated with cleanliness and therefore with holiness, finds wide use at mealtime. For instance, rice is measured in an unfinished wooden measuring box (page 35), served with unfinished wooden paddles (page 35) from unfinished wooden containers (page 34), and eaten, in most cases, with unfinished wooden chopsticks (page 37) together with fish or vegetables that have been cut or chopped on an unfinished wooden board (pages 38–39). The meal may be followed by a cup of tea together with sweets that have been pressed in unfinished wooden molds (page 49) and served from boxes (page 44) or trays (page 45) of the same clean, pure material. Water for the ritual purification of mouth and hands at temples and shrines is scooped from stone basins with ladles of unfinished wood, which is invariably the material from which are made the stands and trays for offerings to the gods. And, having served people well during their lives, wood furnishes the plaques on which to write their names when they die (page 80).

Paper, easily folded, cut, twisted, and oiled, is undoubtedly made to work harder by the Japanese than by any other people. The national passion for wrapping things has given birth to the charming wrapping cloths called *furoshiki* (page 276, bottom), but, in this connection, nothing can take the place of foldable paper. Ordinary wrapping paper is consumed in gargantuan quantities, and the desire to wrap and label gifts has stimulated the development of *noshi* gift markers (pages 122–23 and 156–57), which, in more extravagant forms, may be decorated with sometimes immense and usually quite imaginative *mizuhiki* fancy ties and knots made of twisted paper in gold, silver, and other colors (pages 124–27). The foldability of paper has inspired the growth of the delightful origami craft, which includes such items as love letters and talismans (page 153), human figures for magical purposes (page 155), paper dolls (page 145), decorations for bamboo branches to be displayed for the Star Festival (page 148), and other creations (page 149).

The translucency of paper makes walking with an umbrella less gloomy (pages 136–37) and transforms shōji-fitted windows into walls glowing with a magical light that softens shadows and creates calm (pages 128–35).

Like white wood, white paper, when it serves as part of the decoration on a container for sweetened sakè at New Year's (page 121), symbolizes the purity of a new beginning. Folded into pendants to hang on a branch of the sacred *sakaki* tree (page 160), it signifies a holy presence.

For all their skill in the treatment of metal and ceramics, their reverence for stone, their affection for wood and their indebtedness to it, and their extensive exploitation of the convenience of paper, the Japanese probably reveal themselves most clearly in their ways of dealing with bamboo. This plant, found in abundance in Japan and southern Asia, lends itself to an incredible variety of uses that range from food (the bamboo shoot is a much relished delicacy) to fans and flower vases. It is an architectural material (pages 86–88), a partitioning or curtaining material (pages 81 and 97), and a weaving material for containers of all kinds (pages 102–8). It can be split to become the ribs of umbrellas (pages 110–11), the frameworks for paper lanterns, and the ornamental stoppers for bottles of sakè offered to the gods (page 112). Bamboo makes music (pages 114–16); it provides measuring, spinning, and weaving devices (page 117); and it forms the cages for birds and insects (page 120). It covers hillsides with masses of feathery green plumage, furnishes poles on which to hang the wash, and is in general such a friend to man that one is scarcely surprised to learn that the Japanese consider it one of the four noble plants (along with the chrysanthemum, the plum blossom, and the orchid) and accord it a place of honor in the triad of auspicious plants—pine, bamboo, and plum—unfailingly called upon to dignify things when felicitations are in order or good luck is desired.

FUNCTION AND SYMBOL As has been suggested above, man the maker started out with the most functional applications of natural forms: the stone tied to a stick, the seashell, the large leaf, and so on. Later he copied natural forms in clay and other materials to make dishes to eat from, spears to hunt with, and mats to sleep on. Each object was made for use by one or more persons and filled a particular need only. But as man associated with his fellows as well as with nature, sooner or later everyone in a group decided that a certain form or a certain way of doing things was best. Then the forms assumed independent lives of their own to become generalized and generally accepted. As repositories of knowledge about the techniques required in their production, forms were no longer spontaneous results of creativity but models and guides in imitation. Thus, as arbiters of actions, they came to have a stabilizing effect on ways of life among human beings, who are conservative in most basic matters. In short, instead of making things to perform certain functions, man began to perform certain functions in certain ways dictated by the forms of the implements involved in the act.

In one sense, the conservative forms of generally accepted objects restricted and, in another sense, liberated the craftsman making them. He was bound to turn out dishes that were in consonance with the generally accepted concept of such utensils prevailing among the group in which he lived. But within these limits he was free to decorate or vary the dishes as he pleased. From variations, new forms evolved to meet new needs, and if the new form achieved general recognition among the group, it too became fixed and, in its turn, exerted a stabilizing influence.

But different environmental demands inspired different responses in peoples in various parts of the world, and the same forms did not always achieve universality everywhere. This fact accounts for local individuality in artifacts. Some pot shapes are unmistakably and distinctively Chinese, others just as unmistakably Greek. Archaeologists can date excavation strata on the basis of pottery forms known to have been unique to a given locale or race. In other words, the natures of the forms that became universal and the variations worked on them help to characterize a culture.

When forms become universal within a culture, symbolic meanings accrue to them, and relations between their physical-functional significance and their symbolic significance vary. In some instances the purely physical-functional significance predominates over a very minor symbolic meaning. In other instances the symbolic meaning is as important as the physical one, and in still others the symbolic meaning far outweighs the physical one. The following paragraphs point out some of the forms that have become universal in Japanese culture and attempt to show their ratios of functional and symbolic significance.

Obviously, as long as they are used as originally intended, tools and weapons represent the category in which physical function is more important than symbolism. Though some of their characteristics are distinctive, in general Japanese tools are not unlike those used in many other parts of the world. In addition to fulfilling their defensive and aggressive functions, weapons in Japan can symbolize social status, just as they do in other countries. The armor of a medieval European feudal lord was obviously richer and finer than anything worn by the lowly rank and file. And in Japan only a very high-ranking man indeed would be permitted to bear the burden of the fancy helmet on pages 174–75. But a hammer is a hammer, and a helmet is a helmet, and neither has any great symbolic role to play.

In the case of dishes, function usually determines the form. But the Japanese have specialized eating utensils to a high degree and have given to certain ones generally recognized characteristics that have a kind of symbolism, which is nonetheless secondary to the roles the utensils play in practical matters of serving and eating.

At the bottom of page 26, on the footed tray, stand three bowls. Two of them are practically identical except for size and depth. The other lacks a lid and is different in shape. To the uninitiated it would seem to make little difference what the three are used for, but the Japanese diner knows what to expect in the unlidded bowl (rice) and has a fairly good idea of what will be in the other two before he so much as removes the lids (soup in the deeper one and probably boiled vegetables or something similar in the wider one). In other words, the form in a way symbolizes the content of the dish.

The dashing red and black keg on page 27 is designed not only for sakè but specifically for sakè to be presented as a gift on a special occasion. Though it might seem to be convenient as a server for one of many different kinds of food, the lidded container at the bottom of page 31 is for rice only, and its form announces its function to a person who is in the know.

Before the unisex age, clothing was generally recognized to indicate symbolically the sex of the wearer. This is certainly true of traditional Japanese costumes, and one illustration should suffice to make the point. On pages 70 and 71 are four pairs of geta. A man would ordinarily wear only the kind at the top of page 71. The other three kinds are for women or perhaps for female impersonators appearing on the Kabuki stage.

Some architectural elements and styles illustrate situations in which symbolism exists side by side with practical function but is at the same time definitely subordinate to it. For example, the *azekura* building style, which calls for roughly triangular-section timbers fitted on top of each other and overlapping at the ends in the manner of the American log cabin (page 74), is found only in temple storage buildings (except for the concrete copy of it in the National Theater, Tokyo), and the white clay building with the heavy, often intricately decorated tile roof (pages 202 and 208) can only be a *kura,* or storehouse. Although the style symbolizes the nature of the building, it does not exist apart from the building's actual physical function. The same is true of certain kinds of decorated shōji panels (pages 134–35), which appear only in restaurants or places of entertainment. The untutored foreigner might think it amusing to install them as decorator items in a living room, but for a Japanese to introduce them into a domestic scene would be decidedly suspicious.

The tea ceremony provides many illustrations of objects in which symbolic meaning and practical function are of equal weight. On page 51 is a set of utensils whose forms not only proclaim their functions but also indicate the kind of setting in which they must be used. The red cups on the black stand at top left are for sakè and only for sakè to be drunk at the *kaiseki* meal that often accompanies the tea ceremony. The black container at upper right is for powdered green tea, while the wooden container at bottom is for rinse water, and these utensils are used only in connection with the tea ceremony. All of these objects must function practically. They all have definite roles to perform, but they perform them in a world that is symbolic and nonpractical—that is, the world of the tea ceremony. Consequently, their forms are determined by their functions, but their very functions are symbolic. No one would use the lacquered tea container to hold anything but powdered tea, not even ordinary tea. Although the blue-and-white water container at the bottom of page 222 might look like an attractive bowl for soup, such a use for it would be unthinkable. For it, as for all the other utensils of the tea ceremony, physical function and symbolism have fused. Neither predominates over the other.

There are other things whose functions are from the very outset purely symbolic. The pronged and gilded *kongō* at bottom on page 198 is a replica of a weapon once used in India, but its purpose now, as a Buddhist ritual implement, is the purely symbolic one of destroying defilement in the human heart. In other cases, however, practical functions have figuratively drained away from the object, leaving nothing but symbol. The *noren* door curtain (pages 272–75) may at one time have been physically useful, but today it is all too often physically a nuisance. Nevertheless, it symbolizes the enterprise before whose entrance it hangs more vividly than candy-stripe pole, wooden Indian, or signboard. The *noren* has come to stand not only for the shop but also for the shop's reputation. Misdeeds must be avoided because they bring disgrace to the *noren.* A son or employee who is permitted to set up in business independently with the good name of the parent firm is said to have been given a share of the *noren* and is expected to uphold that good name.

Symbolically speaking, a fan is not always a fan. The heavy-paper gold-decorated dance fans on page 138 are of a kind that once was used against the summer heat, but now, being used strictly for dancing purposes, they have lost their identity and practically their existence in the many things they are called upon to symbolize on the stage: everything from bow and arrow to sakè bottle and sakè cup. The physical functions of the fans have been forgotten. The fans no longer need to do what they look as if they ought to do, for they have been swallowed up entirely in symbol.

Crests are obviously symbols. Like the *noren,* they stand for reputation; like the dance fan, they have a symbolic function only. On page 280 is the triple-hollyhock-leaf crest of the house of Tokugawa, which ruled Japan in relative peace from the seventeenth until the mid-nineteenth century. This crest once inspired awe—possibly even dread—in all who saw it. Today it is an object of interested comment to the people who remember what it means. But not everyone remembers, nor do all Japanese now know the physical or symbolic functions of all the objects shown in these pages. Much has been forgotten, and much is still being forgotten. But the Tokugawa crest and the other objects in this book have something to tell. They speak eloquently to the eye and mind of the concerned viewer. Taken together, they are a myriad-fragment mirror reflecting the way a people lived in the past, continues to live to a limited extent now, but may never live again. The tale they tell is a wordless one of color, form, material, and symbol. It is a rich story of sounds, smells, sights, and emotions. It is too rich indeed to be completely understood from books, although a collection of excellent photographs is a good second best to direct contact with the elements of the story themselves.

Commentaries on the Photographs

by Takeji Iwamiya

This book is an attempt to fix in photographs the forms of some of the objects that the Japanese people of the past devised from natural materials and passed on to us. Although the basic ideas for many of these objects came from China or elsewhere, Japanese hands reworked and altered them to suit the geographic setting, the climate, the customs, and the ways of living in our islands. Many of the things shown in the book originated in the late seventeenth and the early eighteenth centuries, when the increasing wealth of the urban merchant class was undermining the ancient social structure. They therefore reflect the tastes and talents of a vigorously active level of urban society.

Obviously a finished craft product or art object differs from the natural materials of which it is made. But the craftsman's attitude toward his materials exerts a great influence on the results of his work. The Japanese craftsman traditionally believes that he and his materials work in cooperation. Unlike the Western craftsman, who works *on* his materials, the Japanese craftsman prefers to work *with* his materials. He loves to take advantage of the natural qualities and to reveal them. Sensing no duality between man and nature, he allows the colors and textures of his materials to find frank expression in the completed work.

The outcome of this attitude has been a multitude of ordinary objects of extraordinary beauty. In earlier books of mine I have presented photographs of some of these things, but, looking back on those books, I am not entirely satisfied with the selections I made for them. Still, in my preface to one of them I found a statement that is in complete harmony with the opinions I hold today: "In the rapidly developing society of modern Japan, demands for rationalism and functionalism, together with advances in scientific technology, are steadily altering the forms of the things we use. Some of the things our forefathers made and used in times gone by are no longer popular, and some of them have already vanished. Perhaps their loss is an inevitable consequence of the changing times, but I do not want to stand idly by while much that is valuable is being lost."

This book represents an effort to be something more than an idle bystander. My own appreciation of the beauty of traditional forms and the way of life they stand for has developed during the years that have passed since I wrote the statement quoted above. And I believe that for this book my now clearer eye has chosen things that are more beautiful than the ones I selected for earlier books. Here, then, in order that the reader may fully understand the beauty of the objects I have chosen, are my commentaries on the photographs.

Page 25. Footed lacquer tray and dishes for special occasions. This set was used by an old Kyoto family for New Year's celebrations. The deeper bowls are for rice and *miso shiru* (bean-paste soup); the shallower one, for a side dish. The white wrapper, with the Chinese character for long life and the knotted gold-and-silver paper string known as a *mizuhiki,* contains chopsticks. Both the tray and the bowls are adorned with the crane crest of the family. Men's dishes are of red lacquer; women's, of black.

26. Top: footed lacquer tray and dishes for auspicious occasions. Bottom: footed lacquer tray and dishes for offerings at Buddhist ceremonies or at special ceremonies for paying reverence to ancestors.

27. Horned keg, or *tsunodaru,* for sakè. The name derives from the elongated handles, which resemble horns. The inverted cone at center is the stopper. The keg itself is lacquered in red, and the bamboo strips that strengthen it are in black. Kegs of this kind are used at weddings, at ceremonies to celebrate the completion of the framework of a house, and in making offerings to the gods.

28. Tiered food boxes (*jūbako*) of this kind were in wide use from the Muromachi period (1336–1568) through the Edo period (1603–1868) and are still used today to hold delicacies and special treats for the New Year's holidays and the Girls' Day festival on March 3. They are frequently of red or black lacquer and may be richly decorated with patterns in a combination of lacquer and gold powder.

29. Handled tiered boxes (*sageru jūbako*) are designed for carrying lunches on outings to view the cherry blossoms or the autumn leaves. Like the set shown here, they are often richly decorated.

30. Top: lacquer soup bowls. Bowls for soup and other foods may or may not have lids. If they do, the lids may fit within the rim, as they do here, or extend over and below the rim, as do those shown on page 31. The pattern of chrysanthemums and flowing water is one of the most popular in Japanese decorative arts. Bottom: two-part food box in red lacquer repeating the chrysanthemum-and-water design.

31. Top: lidded red-lacquer bowls. Bottom: lidded rice container and rice server in red lacquer. Elegant ware of this kind is used for the *kaiseki* meals associated with the tea ceremony.

32. Set of three lacquer cups for sakè. The grandeur of the pine, the resilience of the bamboo, and the courage of the plum tree (for blooming in the winter cold) have made these three plants favorite decorations for the new year. This set of sakè cups bears the three auspicious symbols.

33. Top view of lacquer pot used for serving sweetened and spiced sakè (*toso*) at New Year's. Like the sakè cups on page 32, it is decorated with emblems of the pine, the bamboo, and the plum.

34. Lidded containers for cooked rice. Though sometimes lacquered, containers of this type are more often of unfinished wood.

35. Top: *shamoji,* or paddlelike servers for rice. *Shamoji* may be of unfinished wood, lacquered wood, or bamboo. Since they were believed to bring good fortune, from the early part of the seventh century they were sold by Buddhist temples and Shinto shrines. Bottom: measuring box (*masu*) for rice and other dry foodstuffs. The system of measuring by *masu* is thought to date from the Nara (646–794) or perhaps the Heian (794–1185) period.

36. Top: *masu* for liquid measure. Each box of this graduated set bears the designation of the amount it contains, the largest being for 1 *shō,* or 1.8 liters. The character for *shō* (升) is also the character for *masu.* Bottom: *masu* for measuring rice and other grains. It is also for 1 *shō,* and the round wooden rod on top is used to level the contents for accuracy.

37. Top: six kinds of chopsticks, or *hashi.* Starting from the top: *waribashi,* or chopsticks that must be broken apart before use; *yanagibashi,* or willow chopsticks; round chopsticks; square chopsticks; Rikyū chopsticks, named for the famous sixteenth-century tea master Sen Rikyū; and lacquered chopsticks. Bottom: husband-and-wife chopsticks in lacquer box (lid not shown). The husband's, by tradition, are the longer ones.

38–39. Footed board (*manaita*) for cutting fish and vegetables. *Manaita* are usually made of cypress, magnolia, or oak. The wooden-handled knife shown here is of steel brought to a swordlike sharpness.

40. Top and bottom: *soroban,* or abacuses. Invented in China, the *soroban* was introduced into Japan between 1592 and 1615, but it has been modified to the extent that today Chinese and Japanese *soroban* are not identical. Nor are they used in precisely the same ways.

41. *Chadansu,* or chest for storing dishes, containers for sweets, and various other utensils used in the tea ceremony. This one is finished in the Kyō-Negoro style of lacquer, in which, as the object is used over the years, the lower layer of black lacquer begins to show through the upper layer of red.

42. Top: *tabako-bon* (literally, "tobacco tray"), or smoking set. Cut tobacco is stored in a small drawer at the bottom. The ceramic pot serves as a brazier for ash and lighted coals with which to ignite the tobacco stuffed into the *kiseru,* or pipe, shown at front. The bamboo container at right serves as an ashtray. Bottom: *nagahibachi,* or long brazier. The name derives from the thick board extending beyond the brazier on all four sides. Various items may be stored in the drawers. The pit for ashes and charcoal is lined with copper, and a footed iron rest and pot for heating water can be set at one end. The metal chopsticks (center) are used for handling hot charcoal, and the metal scoop to their right serves for removing ashes.

43. Board and pieces for the chesslike game called *shōgi.* The game, said to have originated in India, entered Japan by way of China, but the Japanese have modified it almost beyond recognition. The major alteration they have made is to allow a player who captures an opponent's pieces to use them to his own advantage.

44. Top and bottom: wooden boxes for confections. In the upper one a slatted bamboo mat is spread on the bottom, while in the lower one a white paper napkin bordered in festive red has been used to harmonize with the white and pink of the confections. Red and white, when used together, are colors of auspicious connotation.

45. Individual wooden trays for confections. The beauty of the confection is highly important. At top is a round cake with a pine-bough design; at center, one of camellia shape; and at bottom, one in the shape and color of an autumn maple leaf. Both confections and dishes or trays for them occur in an immense range of shapes and kinds to suit all tastes, but traditional Japanese confections fall into two major categories: *namagashi,* or fresh confections, made of a sweet paste of *azuki* beans and therefore soft and perishable, and *higashi,* or dry confections, made of sugar and thus more durable. The ones shown on this page are *namagashi.*

46–47. Black-lacquer tray spread with a variety of *higashi.* The designs include chrysanthemums (top left), holly (top right), autumn maple leaves and mushrooms (bottom left), and camellias (bottom right). These *higashi,* in contrast with those seen on the next page, are made by hand instead of being pressed in molds.

48. *Higashi* made with molds of the type shown on the following page. Molded *higashi* appear in an astonishing variety of shapes: chrysanthemums, camellias, cherry blossoms, oak leaves, pine needles, clouds, and countless others.

49. *Higashi* molds. Molds for preparing these dry confections are usually made of cherry wood.

50. Tiered boxes for tea-ceremony confections. The elegance of these lacquered wooden containers matches that of the confections shown on the preceding pages.

51. Three kinds of utensils used in the tea ceremony. Upper left: a set of five shallow red-lacquer sakè cups resting on a black-lacquer stand. Cups of this kind are employed in the *kaiseki* meal that sometimes accompanies the tea ceremony. Upper right: a container for powdered tea in *natsume* (jujube) style. Although this one is of black lacquer decorated with a gold-lacquer design, *natsume* are frequently of unfinished wood. Bottom: *kensui* basin for rinse water. This one is made of bent wood and is lacquered in red inside.

52. *Biwa,* or lute. Four- and five-stringed lutes from India or regions to the west passed through China to reach Japan during the Nara period (646–794). The *biwa* is played with a plectrum, or *bachi.* Throughout the centuries it developed into a variety of Japanese types: the Raku, the Heike, the Satsuma, and others. In the Meiji era (1868–1912) there appeared the Chikuzen *biwa,* a smaller type about 70 centimeters in length.

53. Koto. The koto, sometimes defined as the Japanese zither, is about 2 meters long, about 15 centimeters wide at the upper edge, and about 24 centimeters wide at the lower edge. It is played by plucking with ivory picks. The strings, generally thirteen in number, run the length of the bowed top, which is of laminated paulownia wood, and are supported by ivory bridges.

54. *Tsuzumi,* or hand drum. Top: drumhead; bottom: side view. Although they originated in China, drums of

this kind were perfected in Japan. The *tsuzumi* is held on the left shoulder by the left hand, which adjusts the cords for sound effects, and is struck with the right hand.

55. *Taiko,* or large drum. The *taiko,* made in a variety of sizes, is struck with a drumstick of leather-covered wood.

56. Shamisen. The three-stringed banjolike shamisen, which developed around the middle of the sixteenth century, derives from the Ryukyuan *jabisen.* It reached its final stage of evolution in the Edo period (1603–1868). Consisting of a sound box covered with animal skin (often snake or cat), a long neck, strings, a bridge, and pegs for adjusting the tautness of the strings, it is plucked with a large wood-and-ivory plectrum and enjoys great popularity in furnishing the accompaniment for traditional songs and for the Kabuki and puppet dramas.

57. Noh mask. This type of mask, known as *ko omote,* is used for the roles of beautiful young women.

58–59. Noh masks. The distinctive masks employed in the Noh drama developed over a long period under the influence of more ancient dramatic forms like Gigaku and Sarugaku. Characters who appear in the Noh may be classified into five basic groups: gods, ghosts, beautiful women, mad people, and demons. There are thus five basic types of masks corresponding to these categories, but variations resulting from the preferences of the different acting schools bring the total to more than a hundred. The masks on pages 58 and 59 are, at top: *masukami,* a young woman possessed by spirits; *jōjō,* a smiling, red-faced, drunken person; *okina,* a venerable old man; and *shikami,* a scowling, malevolent man; and, at bottom: *deigan,* a female ghost with golden eyes; *chūjō,* an aristocratic young man; *hannya,* a female demon; and *ōbeshimi,* a sorrowful male demon.

60. Bunraku puppet head. Each of the major puppets in the traditional Bunraku (puppet drama) is operated by three men. The doll consists of a head, a trunk, arms, and legs. The eyes (and sometimes even the eyebrows), the mouths, and the joints of the hands can be moved by the operators to heighten dramatic effectiveness. Bunraku originated in the Edo period (1603–1868) and was greatly developed by Takemoto Gidayū (1651–1714), who produced dramas to be performed to the accompaniment of a thick-necked version of the shamisen.

61. Two Bunraku puppets: a playboy and his courtesan sweetheart. The hand of one of the operators is partly visible behind the right hand of the playboy.

62–63. Bunraku puppet heads. The heads, reading across this two-page spread, are, at top: a middle-aged woman, a young girl, a young girl of high rank, and a young man; and, at bottom: a young boy, a middle-aged man (the Danshichi head), another middle-aged man (the Bunshichi head), and another version of the Bunshichi head.

64. Left: *hagoita,* or battledore, for the game of battledore and shuttlecock, traditionally played at New Year's. *Hagoita* for ornamental purposes are often of tremendous size, and the practice of decorating them with raised figures of famous people (in this case, Sukeroku, the hero of the Kabuki play named for him) by using cardboard, cotton padding, and a silk covering is thought to have developed in the period between 1804 and 1830. Right: wooden tops. The game of tops reached Japan from China by way of Korea before the eighth century.

65. Regional toys of wood. Upper left: a top from the Tōhoku region of northern Honshu that spins on the head of a wooden figure of the Buddhist patriarch Daruma (Bodhidharma). Lower left: a wooden horse from Miharu, in Fukushima Prefecture, traditionally thought to have a good influence on the rearing of children. Right: a *kokeshi* doll from the Tōhoku district. *Kokeshi* dolls are made in a wide variety of shapes by turning wood on a lathe. Most of them have round heads like that of the young girl shown here.

66–67. Palace dolls for Girls' Day. Dolls of this type, traditionally costumed in the court style of the Heian period (794–1185), are made in a number of styles and sizes. They are displayed by families with small daughters on March 3. A full set of dolls may include, in addition to the ruler and his lady, many courtiers and ladies-in-waiting as well as miniature articles of furniture and even food and drink, all arranged on a tiered platform covered with a bright-red cloth. According to a tradition that is still older than that of displaying such dolls, paper figures of ancient palace personalities are floated on streams in celebration of the holiday. This custom is thought to have originated in Heian times.

68–69. Boxwood (*tsuge*) combs. Boxwood combs for hairdressing and for ornamenting the feminine coiffure have been popular since the Heian age. On page 68, from top to bottom, appear the coarse comb, the flowing comb, the square-and-round comb, and the Rikyū comb. On page 69, from left to right, are a comb for smoothing out separations, the Shinagawa comb, the five-toothed comb, the Miyako, the combination comb-spatula, the halberd comb, another Shinagawa comb, the Genroku comb, the rat's-tail comb, and the clam comb.

70–71. Geta, or wooden clogs. It is thought that the origin of geta can be traced to the wooden footgear worn in the fields during the prehistoric Yayoi period (200 B.C. to A.D. 300). In the Edo period, geta of paulownia wood, which is both strong and light, became popular. Page 70 shows women's geta; page 71, top, men's; page 71, bottom, women's. The lacquered ones are for formal wear.

72. Post-mounted pincushion (left and lower right) and thread on wooden bobbin (upper right). The piece of fabric being sewed is attached to the cord on the post for easy manipulation.

73. Lacquered kimono rack. This one is in two sections, hinged at center.

74. Corner of storehouse building in *azekura* style. This ancient architectural style, which resembles that of the American log cabin except for using triangular timbers, dates from the Nara period (646–794). The example shown here is part of the Nigatsudō complex at the Nara temple Tōdai-ji. Other famous and ancient surviving examples are the Shōsō-in treasure repository at the Tōdai-ji and the storage buildings at another Nara temple, the Tōshōdai-ji.

75. Base of corridor post, Hōryū-ji. The post is one of those in the corridor surrounding the main compound of the seventh-century temple Hōryū-ji, in Nara, and is set on a foundation stone. The striped pattern on the corridor floor results from shadows cast by sunlight falling through the slatted corridor windows.

76–77. Miniature wooden torii at the Fushimi Inari Shrine in Kyoto. Small versions of traditional gates to Shinto shrines are offered to the Inari fox deities, who are thought to bring luck to commercial ventures.

78. *Hishaku,* or wooden water ladles. Traditional utensils like these are still in wide use and are much preferred to their metal counterparts.

79. Footed ceremonial tray (*sambō*). Although in ancient times trays of this type were used in serving meals, today they are restricted to offerings to Shinto gods or to Buddhas. The *sambō* shown here, made of unfinished cypress, rests on a base with a perforated decoration and holds *dango,* or rice dumplings.

80. *Ihai,* or memorial plaques. *Ihai,* bearing the posthumous names of the deceased, play a central role in Buddhist funerals and memorial services.

81. *Sudare,* or bamboo blind, at the Nara temple Hokke-ji. Elaborate *sudare* of this type, made of the finest split bamboo stitched together and bound with brocade or satin, are used today only in shrines and temples. The blind may be raised and held in place by means of the tasseled cord and the metal hook.

82–83. Section of bamboo fence at Katsura Detached Palace, Kyoto. This fence, named for the imperial villa where it stands, consists of vertically set split sections of bamboo with diagonally cut upper tips and of horizontally placed bamboo twigs tied in place with cords.

84. Kōetsu-style fence at the temple Kōetsu-ji, Takagamine, Kyoto. The unusual style of fence seen here is named for the famous lacquer artist and calligrapher Hon'ami Kōetsu (1558–1637), for whom the temple is also named. The style employs a combination of a rhombic lattice of heavy bamboo and a coping of split and bundled bamboo.

85. *Yarai,* or palisade-style fence, at the Kyoto temple Daitoku-ji. In the past, fences of this kind served a protective purpose. This one tops a stone wall at the Kohō-an subtemple of the Daitoku-ji.

86. Circular window designed to suggest that the latticing is actually the wattles on which the clay wall is laid. Windows of this kind were widely used for rooms in which the tea ceremony was performed. The one shown here is in the Kōbai-in of the Daitoku-ji, in Kyoto.

87. Octagonal window aesthetically exposing the wattles of the clay wall. Kohō-an, Daitoku-ji, Kyoto.

88. Window and *nijiriguchi* (guest entrance) of Tōyōbō teahouse at the Kyoto temple Kennin-ji. To emphasize their equality in the tea ceremony, the guests must enter the tearoom through a low door that requires them to crouch.

89. Tea-ceremony utensils on tatami mat. Water is heated in the iron kettle on the brazier at left. Water from the lidded vessel at right is ladled into the kettle to adjust the temperature of the tea water in the kettle. Powdered tea is taken from the lacquer container with the scoop that rests on top of it, and the bamboo whisk is used to mix the powdered tea with boiling water from the kettle.

90. Top: green-bamboo chopsticks for use in the *kaiseki* meal that sometimes accompanies the tea ceremony. The chopsticks are classified by shape and the location of the natural joint—hence, top to bottom: middle joint, tapered at both ends, and end joint. Bottom: whisks (*chasen*) used in preparing ceremonial tea.

91. Bamboo basket for charcoal and implements for placing it in the tea-ceremony brazier: feather brush, ring handles for removing the brazier, metal chopsticks for handling the charcoal, black charcoal, white charcoal (for aesthetic purposes some schools of the tea ceremony coat branches of ordinary charcoal with Chinese white), and ceramic rest for lid of kettle.

92–93. Hanging bamboo containers for tea-ceremony flower arrangements (*chabana*). Flowers like the camellia on page 92 and the clematis on page 93 are preferred for the tea ceremony.

94–95. Containers for tea-ceremony flowers may be made of simple sections of bamboo with openings cut in various positions, as on page 94, or they may be woven bamboo baskets like the one at left on page 95. At right on page 95 is a basket with a cloth drawstring top for carrying tea-ceremony implements of small size.

96. Entranceway flower arrangement in bamboo container at the Obiya, in Kyoto. A simple length of green bamboo holds New Year's flowers and offers a warm welcome to guests. It also harmonizes beautifully with the traditional *minka*-style architecture—that is, the architecture of the essential Japanese house in premodern times.

97. In Kyoto, as in other places in Japan, *noren* entranceway curtains are usually of cloth. At this *minka*-style residence near the Daitoku-ji, however, the *noren* is of bamboo.

98–99. Festival lanterns of paper and bamboo like the ones seen here are hung under the eaves at front doors to contribute to a general holiday mood. Sometimes they bear the names of commercial enterprises housed in the buildings where they are hung.

100. Portable paper lantern. When it is stretched to full size, it is held in position by the ring that hooks over the bamboo bow.

101. Gifu lantern. The ribs of Gifu lanterns are especially fine, and the covering of cool-colored paper is decorated with paintings of flowers and grasses. Suspended from the eaves, these lanterns, which are made largely in Gifu Prefecture, give a pleasant illusion of relief from summer heat. Since they are often placed in front of Buddhist shrines during the summertime Bon festival in remembrance of the dead, they are also known as Bon lanterns.

102. Top: summer container for cooked rice. *Hachiku* bamboo, of which these basketlike containers are made, lends itself to shaving and fine weaving. Bottom: simple basket for washing grain before cooking.

103. Top: basket with high handle for drying fish. The open weave permits good circulation of air. Bottom: draining basket for vegetables, fruit, and other items washed in the kitchen.

104. Decorative basket of finely split and intricately woven bamboo. The hexagonal pattern at center suggests the shell of a tortoise and is a favorite in Japanese decorative design.

105. Top: bamboo sieve. Bottom: bamboo winnow for rice and other grains.

106. Deep basket for picking of vegetables like beans and peas. Both this basket and the one on page 107 display the interesting decorative use of green and aged bamboo.

107. Deep handleless basket of green and aged bamboo. It has a variety of household and field uses, and its pleasing shape demonstrates the skill of the craftsman.

108. Top: creel of traditional shape. Bottom: deep lidded basket for holding eels.

109. Garden rakes and open-ended basket for fallen leaves. The Japanese call such rakes *kumade,* or bear hands.

110–11. Traditional oiled-paper umbrella (*karakasa*) of the type known as *janome,* or snake eye. See also pages 136–37.

112. Decorative bamboo stoppers for bottles of sakè offered to the gods. The delicate workmanship speaks for itself. Small squares of red paper are pasted over the overlapping joints.

113. *Fude,* or writing brushes, for use with *sumi* ink. They are in the traditional style of those imported from China when writing was introduced into Japan.

114. *Shō,* a type of wind instrument introduced from China in the early seventh century and used to accompany performances of the court music and dance known as Gagaku. The *shō* consists of seventeen bamboo pipes rising from a kind of bowl. Two of the pipes produce no sound. Tonal gradations are made by fingering the holes, and air is inhaled through the large hole in the base.

115. *Yokobue,* or transverse flute, a fifelike wind instrument used to accompany a variety of traditional performing arts. It is held in the same style as the modern Western flute and has seven finger holes.

116. *Shakuhachi,* or bamboo flute. The *shakuhachi* was imported from China by way of Korea in the mid-seventh century. The name derives from the accepted standard length: one *shaku,* or 30.3 centimeters, plus eight (*hachi*) *sun,* a *sun* being equal to 3.03 centimeters. The instrument plays in fifths and has four finger holes.

117. Upper left: *itomaki,* or reel for spinning thread. Lower left: *hi,* or shuttles. Right: *monosashi,* or rulers. The ruler, a very ancient measuring device, is said to have been used in ancient India some three thousand years ago. Specifications for Japanese rulers appear in the Taihō Code of 701.

118. New Year's ornament in shape of *takarabune,* or treasure ship. Except for the bamboo-and-pine mast and the cords, it is made of rice straw and has heads of rice for the prow. Just above the sail are artificial plum blossoms and bamboo leaves, which, together with the pine, make up the group of plants considered suitable for auspicious occasions. Ornaments of this kind are placed on *sambō* trays like the one shown on page 79.

119. New Year's ritual straw rope and seasonal offerings of red and white nandina, Oriental bittersweet, chrysanthemums, ornamental purple and white cabbage, and pine branches, all of which are traditional decorations for this most important of all Japanese holidays. Ritual ropes are usually hung in sacred places to divide the holy from the secular and the unclean. For this reason they are always made of completely fresh materials.

120. Top: *torikago,* or bird cage, of bamboo. Bottom: bamboo insect cage (*mushikago*) for bell crickets and other insects kept as pets.

121. Folded red and white handmade paper decorated with the traditional felicitous emblems *shōchikubai*—pine, bamboo, and plum blossoms—and with a *mizuhiki,* or knotted paper cord, of gold and silver and placed on top of a lacquer pot for the spiced and sweetened sakè (*toso*) drunk at New Year's to ward off evil.

122–27. *Mizuhiki* and *noshi.* Both of these serve as ornaments for special occasions. *Mizuhiki,* made of twisted paper cords stiffened with a starch-and-water solution, decorate gifts and utensils. At festive times they are red and white, gold and silver, or gold and white. For sorrowful occasions they are black and white or blue and white. For weddings the paper cords are intricately worked into such auspicious symbols as the crane (page 126) and the bushy-tailed tortoise (page 127, bottom), both of which mean good luck; the treasure ship, which has obvious connotations; or pine, bamboo, and plum. Page 127, top, shows the underside of a *mizuhiki* in the shape of a butterfly. In the distant past, strips of abalone were used to signify that an object was a gift and were wrapped in square pieces of colored paper folded into irregular hexagons—wider at the top than at the bottom—like the ones on page 123. In later times, to make these *noshi,* as they are called, the abalone was replaced with a piece of paper attached to another piece that was folded and more or less elaborately decorated and used as a label to indicate the occasion for which the gift was offered. The *noshi* sometimes bore the name of the giver. Simpler *noshi* labels appear on pages 122 and 123. The decoration, much more elaborate in the *mizuhiki* on page 125, reaches a peak of ornateness in the one on page 124. In all cases the strip of abalone is represented by an amber or yellowish strip of paper.

128. Shōji, or sliding panels of paper on wooden frames, serving as doors or windows. The ones seen here are in a traditional *minka* (commoner's house) in Kyoto.

129. Shōji at the temple Hokke-ji, in Nara. The decoration is a crest showing part of a chrysanthemum.

130. Shōji at the famous temple Kiyomizu-dera, in Kyoto.

131. Doors with sections of shōji at top, Hōrin-ji, Kyoto. The drawings are of Bodhidharma (in Japanese, Daruma), and the Hōrin-ji is popularly known as the Daruma temple.

132–35. Shōji in the Sumiya, in Shimabara, an area once officially licensed as the pleasure quarters of Kyoto. The Sumiya is the only remaining wooden two-story building that represents the old-time *ageya,* or place to which geisha were summoned to sing and dance or otherwise entertain. Geisha did not actually live in *ageya*. The Sumiya was designated an important national cultural property in 1962. Originality and ingenuity distinguish the designs of its shōji, which are different in each room.

136–37. *Janome,* or snake-eye, paper umbrellas. Popularly used since the late seventeenth century, the *janome* umbrella is so called because, when it is open, it resembles the eye of a snake. The umbrellas may be of black and white, red and white, or indigo and white. See also pages 110–11.

138. *Mai ōgi,* or dance fans. The folding fan was invented in Japan during the Heian period (794–1185). The richly decorated gold-paper-covered ones shown here are not for use against the heat but for countless subtle expressive functions in the traditional dance.

139–41. *Uchiwa,* or flat fans. The *uchiwa* was introduced from China into Japan in ancient times. It is made by splitting bamboo into fine ribs, leaving just enough of the stalk intact to form a handle, and then covering the framework with paper or silk. Many different kinds and shapes of *uchiwa* are in use today. Page 139 shows a festival fan, while the one at top on page 140 bears a crest. The red fan at bottom on page 140 is decorated with a coin labeled "one million *ryō*" (a *ryō* was a feudal-period coin of considerable value) and a magic mallet that, when wielded by Daikoku, god of wealth, can make it rain money. The *uchiwa* on page 141 is covered with delicately cut red paper that leaves the ribs partly exposed.

142–43. *Tako,* or kites. The kite, imported from China, was popular with aristocrats and samurai during the Kamakura period (1185–1336), but during the Edo period (1603–1868) it was enjoyed mostly by ordinary people. Kites come in many forms and are decorated in countless ways. The one on page 142 carries a picture of the mythical empress Jingū and the equally mythical warrior Takenouchi Sukune, who is said to have saved Jingū during a campaign to conquer the Korean kingdom of Silla and to have served five emperors over a period of 244 years. Page 143 shows an Enshū kite with a triple comma (*mitsudome*) pattern, a pattern of rhombuses, and a dance fan of the felicitous type used in the Noh drama.

144. Palace dolls of paper and cloth, with wooden heads. The costume design includes the crane along with the pine, the bamboo, and the blossoming plum tree. The venerable couple at the bottom of the female doll's costume symbolize a long and happy marriage.

145. Palace dolls of folded paper. The more elaborate dolls for Girls' Day appear on pages 66 and 67. The ones shown here are of far simpler type. The custom of making paper replicas of palace dolls originated in a beautiful and ancient custom. Tiny dolls in their own small boats, made of straw (top) or wood (bottom), are set afloat on streams or on the sea as prayers for safety and protection from harm.

146–47. *Koinobori,* or carp banners. On May 5, formerly Boys' Day and now Children's Day, families with young sons display several *koinobori* on flagpoles topped with wheels made of representations of arrow feathers. When the wind blows, it enhances the gaiety of these festive decorations. The custom originated in the mid-Edo period.

148. Ornaments for Tanabata, or the Star Festival, celebrated either on July 7 or on August 7, according

to whether the modern calendar or the old lunar calendar is followed. For this occasion fresh boughs of bamboo decorated with strips bearing poems or prayers for progress in calligraphy (the written sheets shown here) or in sewing (the miniature paper kimono) are displayed in front of residences and shops. The Tanabata celebrations in Fujisawa (July 7) and Sendai (August 7) are especially famous for lavish display.

149. Origami creations. At top, left to right, are a *hakama* (traditional skirtlike-trouser garment), a feudal-period serving man, and a helmet. At bottom are three *kazaguruma,* or pinwheels.

150–51. *Kirigami,* or paper cutouts. *Kirigami* of this type are pasted on family Buddhist shrines or household lintels at New Year's and on the columns and eaves of Shinto shrines at various festival times. The one on page 150 carries the characters for "great fish." On page 151, at top left, are a tortoise and a crane and, at top right, a treasure ship. At bottom on the same page is a carp, a fish associated with happy occasions.

152. Shinto protective talisman (left) and ritual paper ornaments. Talismans like this one from the Yasaka Shrine in Kyoto are issued by Shinto shrines and Buddhist temples to ensure protection from harm. On the cluster of *chimaki*—a festive food made by steaming glutinous rice in bamboo leaves—are red-and-white paper ornaments of the type used by Shinto shrines. Such ornaments may be red and white, all white, gold and silver, or a combination of five colors. The *chimaki* shown here were made for the famous Gion Festival in Kyoto: an annual summertime celebration.

153. Talismans folded in traditional love-letter style. The one at far right is decorated with a paper plum blossom.

154. Protective talismans from the Kasuga Shrine, Nara (left); the temple Todai-ji, in Nara (center); and the temple Kiyomizu-dera, in Kyoto.

155. Paper images in human form. They are used in ritual purification ceremonies.

156–57. Paper symbols of gratitude offered on visits to a Shinto shrine. The ones seen on these two pages were presented by parents to the Yasaka Shrine, in Kyoto, on the first visit following the birth of a child.

158. Paper talismans pasted at the entrance to the Sumiya (see pages 132–35). The face is that of a god. The talismans, all of which are from the Sumiyoshi Shrine, are believed to ensure domestic security and to ward off disease.

159. *Omikuji,* or fortune papers. Shrine or temple visitors pray to the gods or Buddhas and then buy papers like these to learn what good or evil lies in store for them. The papers are not carelessly discarded. In this case they have been thrust on spikes or knotted around them. At some shrines and temples they are tied to the branches of trees.

160. *Gohei,* or ritual Shinto paper ornament, on bough of sacred *sakaki* tree. This decoration is attached to the south gate of the Yasaka Shrine in Kyoto.

161. *Kugikakushi* (literally, "nail concealer"). Strength and dignity characterize the design of this nail cover on the bell tower of the tenth-century temple Byōdō-in, at Uji, near Kyoto.

162–63. Finger-grip door pulls on interior sliding doors (*fusuma*) at the temple Hokke-ji, in Nara. The vermilion tassels and the floral motif of the metalwork are suitable to a temple for Buddhist nuns. The tassels are purely ornamental, for the *fusuma* are opened by inserting the fingers in the metal pulls.

164–65. Hanging bronze lanterns at the Kasuga Shrine, Nara. An immense number of these lanterns, with many different patterns in the filigree of their light chambers, hang from the eaves of the red-and-green corridors of the main hall of this celebrated shrine. When they are all lighted after dark, they create a mood of mystical loveliness.

166. Hexagonal hanging lantern of bronze, Nara National Museum. The walls of the light chamber are decorated with pierced designs of plum blossoms, cherry blossoms, wood sorrel, mandarin-orange blossoms with water and water plantain, chrysanthemums with a woven fence, and pine and bamboo.

167. Hanging iron kettle for the tea ceremony. Kettles often rest on pronged supports set directly into the hearth, but this cylindrical one is suspended from the ceiling. The inside of the sunken hearth is finished in the same clay as that used for the tearoom walls, and the white branches mixed with the charcoal are actually

charcoal coated with Chinese white for aesthetic reasons.

168. Cast-iron kettle for tea ceremony. This type of kettle goes by the name of *araregama* (hail kettle) because of the tiny knobs that decorate it. *Araregama* may be either large or small, and the pattern may cover all or only part of the surface.

169. Hanging cast-iron tea-ceremony kettle with bronze lid. It is in the style known as crane's neck.

170. Cast-iron tea-ceremony kettle with bronze lid. This kind rests directly on top of a special metal brazier for the charcoal.

171. Spouted cast-iron kettle with bail handle. The style is said to have evolved from ordinary tea-ceremony kettles like those shown on the preceding pages. Spouted kettles are cast in Kyoto, Osaka, and Morioka.

172. Three pairs of metal chopsticks for handling charcoal in hearths and braziers. They are generally made of iron or bronze and may be plain or ornamented. Some have wooden "handles."

173. *Teshoku,* or portable candlestick, for use at night-time tea ceremonies or other nocturnal gatherings.

174–75. *Kabuto,* or feudal-period helmet (two views). The evolution of the helmet to this elaborate state required many centuries. The pronglike projections over the crown, known as *kuwagata,* or hoe shape, and the rampant dragon suggest the warrior's proud lineage. The metalwork and the intricate lacings attest to the makers' craft. A lacquered mask was often worn below the helmet to protect the warrior's face.

176. Upper section of sword blade. The Japanese sword, made of patiently tempered and polished steel, reached a level of technical perfection and beauty unsurpassed throughout the world. More than one hundred swords are included in Japan's national art treasures.

177. *Tsuba,* or sword guards, of pierced iron. The *tsuba,* placed between the hilt and the blade, prevented the opponent's blade from sliding downward to injure the swordsman's hand or arm. *Tsuba* are usually of iron and may be decorated with piercing, as in these examples, or with carving, inlay, or gilding.

178. Detail of *tansu* (chest of drawers) showing iron decorations. In northern and northeastern Honshu, chests for storing small tools and other equipment are bound in iron and elaborately decorated with iron plates, locks, and pulls of great beauty.

179. Lock on carved and gilded doors at the famous Nikko shrine Tōshō-gū. The lock, comprising a shaft and locking device thrust through rings mounted on the door, is of a kind imported long ago from China. By the Edo period, locks had come to be considered fit objects on which to lavish decorating skills.

180–81. Cutting tools for everyday use. On page 180 appear a knife for cutting leather or paper (top left), ikebana (flower arrangement) shears (top right), and three pairs of scissors used in sewing (bottom). Page 181 shows a pair of garden shears.

182–83. Carpenter's tools. Left to right: hatchet, chisel, plane, awl, hammer, saw. The Japanese saw is used with a pulling rather than a pushing motion.

184–85. Angle rule and some of the many planes used for a wide variety of purposes. The angle rule is marked in the old-style measurement system of 10 *bu* to 1 *sun.* A *sun* equals 3.03 centimeters.

186–87. *Sumitsubo* and an assortment of chisels (*nomi*). Although it is made of wood, the *sumitsubo* (literally, "ink pot") at the top of page 186 is at home in this section of the book because it belongs with the general line of tools illustrated here. To make straight lines with a *sumitsubo,* cord in the right-hand part is unreeled through the India ink in the well at left and extended for the length required. It is then drawn taut and snapped to produce a black line. Except for the difference between black and white, the *sumitsubo* is much like the Western carpenter's chalk line. The chisels shown here, like their Western counterparts, vary in shape and size according to their use.

188. Two types of woodcutter's hatchets. The sturdy and highly durable heads are firmly attached to the handles.

189. Top left: small crowbar. Top right: nail puller. Bottom: a variety of awls and drills.

190–91. An assortment of *nokogiri,* or saws, with a variety of teeth reflecting a great variety of uses. As noted above, the Japanese saw cuts on the pull.

192. Top: gardener's shears for pruning. Bottom: a variety of trowels (*kote*) used in plastering, bricklaying, cement work, and other occupations.

193. *Kama,* or sickle. In Japan the traditional implement for harvesting grain or cutting tall grass is the *kama* rather than the scythe, but harvesting by machine has almost completely supplanted its commercial use.

194–95. Top left: women's hair ornaments in traditional style. In ancient times, Japanese women adorned their hair with flowers and grasses, but in the Edo period combs and fancy pins of metal, lacquered wood, ivory, and tortoise shell came into fashion and have remained popular for the traditional headdress worn with the traditional kimono on such occasions as New Year's. Pins like the ones shown here, frequently floral in motif, were decorated with gold, silver, coral, crystal, agate, jade, and other precious or semiprecious metals and stones. In some instances—notably in the headdress of the daughters of wealthy lords and of high-ranking courtesans in the pleasure quarters—the number and elaboration of such ornaments became staggering. Top right: *kiseru,* or pipes. Used by men and women alike in the past, *kiseru* usually have metal bowls and bamboo shafts. A small amount of tobacco was pressed into the tiny bowl, lighted, and puffed until gone. The process did not take long, and today, when the cigarette reigns supreme in Japan, the *kiseru* is out of fashion. Bottom: scale for measuring weight. The object to be weighed is placed on the hook at left. A series of weights, like the one hung from the pole on the right, is kept at hand. The weights are changed until the correct balance is found.

196–97. Cutlery. The knives shown here, known as *hōchō,* are part of a countless variety, the shape of the knife varying with the kind of foodstuff being cut.

198–99. Ritual objects for a Buddhist altar: candlestick; *kongō* (Sanskrit: *vajra,* or thunderbolt), which symbolizes the destruction of evil; vase for flowers; incense burner; and another candlestick.

200. Buddhist priest's staff, of which only the top is seen here. The shape of the metal top is based on that of a stupa reliquary, and the loose rings produce a jingling sound when the staff is shaken. This ritual implement is sometimes used to beat time during the chanting of sutras.

201. Section of farmhouse tile roof. Tile roofs were introduced into Japan from Korea and China during the Asuka period (552–646). The tiles seen here are of the type called *sangawara,* which has one or two of the corners tripped off.

202. *Kura* (storehouse) in the Hōkō-in of the Nara temple Yakushi-ji. Most traditional *kura* in Japan employ a wooden basic structure with a wall base of round bamboo and palm fiber and a thick covering layer (20 centimeters) of clay over which there is a finishing layer of plaster.

203. Window and lower section of wall of *kura* at the Nara shrine Ikoma Seiten. The patterned part of the wall is in the *namako,* or sea-cucumber, style. The effect is produced by inserting tile plates in the wet clay wall and allowing plaster to fill the spaces among them. The style originated in the Edo period and became quite popular during the Meiji era (1868–1912).

204–5. A variation of the *namako*-style wall. This one, at the Kyoto temple Higashi Hongan-ji, is about 100 meters long.

206–7. Eaves tiles at the Nara temple Tōshōdai-ji. Round tiles like these are sometimes decorated with lotus, chrysanthemum, or other patterns, but these bear the name of the temple inside a beaded circle.

208. Gable decorations in temple style. Here the round tiles are ornamented with stylized chrysanthemums. The massive ridge-end decoration in roofing-tile clay at the top is called a *shishiguchi,* or lion's mouth, and the fancy white decoration under the gable is known as a *gegyo,* or pendant fish.

209. Top: demon tiles, or *onigawara*. These are used to grace roofs and, more practically, to stop the leakage of rain at the ends of the main roof ridge and of the descending subordinate side ridges. Horned demons became popular subjects for these tiles after the Muromachi period ended in 1568. Bottom: round eaves tiles. The one at left is decorated with beading and the ancient triple-comma pattern; the one at right, with a lotus blossom.

210. Teapot (*dobin*) of white porcelain with blue fishnet pattern. The teapot, indispensable to daily life in Japan, is thought to have evolved from medicine

containers and the type of iron kettle seen on pages 167–71.

211. Top: small teacups, teapot, and lipped bowl for cooling hot water to proper temperature for tea of finest flavor. Ceramics of this kind, especially Banko and Tokoname wares, are popular for a form of the tea ceremony. Bottom: top views of two porcelain teapots. Pots like these are found in an immense variety of shapes and patterns and with many different kinds of handles.

212. Top: two ordinary cups for the numerous daily drinks of tea that most Japanese consume. Bottom: two ordinary lidded porcelain rice bowls.

213. Ceramic mortar (*suribachi*) and wooden pestle for kitchen use. Bowls of this kind date as far back as the Heian period. Linear grooves are cut into the clay in several directions before firing. The pestle, for which willow, mulberry, and mountain ash are considered the best materials, is scraped over the grooves to grind grain, several kinds of yams, sesame seeds, and other foodstuffs.

214. Shigaraki pot. The town of Shigaraki, in Shiga Prefecture, gives its name to the ware that has made it famous. The kilns located there are considered to be among the oldest in Japan.

215. Salt pot. Containers of this kind, always glazed and lidded, are found in kitchens all over the country. Most of them are similar in shape and size.

216. Top: sakè cup and sakè bottle (*tokkuri*). Such highly popular ceramic pieces as these are made in many sizes, shapes, and colors. Bottom: *tokkuri* of Bizen ware (left) from Okayama Prefecture and of Ōhi ware (right) from Ishikawa Prefecture.

217. *Katakuchi,* or lipped bowl. Undecorated and simple in form, bowls of this type are made in many parts of Japan and are used for pouring oil, sakè, or soy sauce into narrow-mouthed containers. Bottom: *yukihira,* or lipped and lidded pot for preparing soft rice gruel. The container takes its name from that of a handsome Heian-period aristocrat, Arihara no Yukihira (818–93), who, legend says, was once forced by a woman diver to fetch sea water and evaporate it to make salt—presumably in such a pot.

218. Top view of tea bowl of red Raku ware containing frothy green ceremonial tea and standing on a tatami mat.

219. Four tea-ceremony bowls (*matcha chawan*). Top: Hagi ware, deep and shallow types. Bottom left: Oribe ware. Bottom right: Akae ware, so named because of its decoration (*akae* = red picture).

220. Food containers for *kaiseki* meal sometimes served in conjunction with the tea ceremony. These dishes are called *mukōzuke,* or "over there" dishes, because they are placed on the far side of the tray as it stands before the diner. They are found in many shapes. The ones seen here have an underglaze pattern in blue.

221. Left: unglazed ceramic incense container in shape of bell. Containers of this kind may be of pottery or wood and are made in many shapes. Incense is frequently burned during the tea ceremony. Right: incense burner. Burners like these, originally used in Buddhist rituals, may be ceramic or metal and of practically any shape suitable to their function.

222. Top: powdered-tea container of Bizen ware with ivory lid. Small jars like these are treasured by tea-ceremony devotees and are usually brought into the tearoom in rich brocade covers. Bottom: lidded underglaze-blue container for water used to bring the hot water in the tea-ceremony kettle to the proper temperature for brewing tea. Containers of this type, called *mizusashi,* may be of pottery, metal, wood, or bamboo and may have a lacquered lid, as in this case, or a lid of some other material.

223. Bizen-ware vase for tea-ceremony flowers. Bizen is famous in Japanese ceramics as one of the Six Ancient Kilns, which date from the Kamakura period, and its wares are characterized by a sense of strength and long tradition.

224. White porcelain hibachi (brazier) for live charcoal to warm the hands or to heat water for tea. Hibachi come in many different shapes, sizes, and types and may be made of wood, metal, or pottery.

225. Miniature Daruma (Bodhidharma). Unlike the beetle-browed, bearded papier-mâché figures of Daruma often kept as talismans for success in business

or for a good harvest, the miniature ceramic ones made in the Shikoku prefectures Ehime and Kōchi are quite charming and appealing in appearance—like the one shown here.

226–27. Fushimi figurines made and sold as souvenirs at the Fushimi Inari Shrine, in Kyoto. These unglazed and painted ceramic figures come in hundreds of shapes, including those of the gods of wealth and good fortune, Ebisu (page 226) and Daikoku (page 227).

228–29. Painted pottery pigeons and wooden votive plaques (*ema*) are offered at the Miyake Hachiman Shrine, in Kyoto, to exorcise children's illnesses. The god honored by the shrine is the emperor Ōjin (reigned 270–310), who is responsible for martial courage and is therefore thought to have sufficient power to drive away the demons of disease.

230. Ornamental sakè bottles for presentation to the Shinto gods at the Kitano Temman Shrine, in Kyoto.

231. Vessels for the rice offered to the gods at the Kyoto Kitano Temman Shrine along with the sakè.

232. Unglazed triple cups with central cloud pattern. These cups, shown resting on a tray of unfinished wood, are used for presentation of sakè to the Shinto gods or for actually drinking sakè on special ritual occasions.

233. Stone basin at the Kyoto temple Ryōan-ji. Since ancient times it has been customary to wash the hands before approaching Shinto and Buddhist places of worship. In earlier days, natural stones, metal basins, or wooden tubs were used, but in the Kamakura period carved stone basins like this one became popular. The characters around the square well mean "It is sufficient to know oneself."

234. Steppingstones and pond in stroll garden of the Heian Shine, Kyoto. The stones were once the tops of old granite columns.

235. Steppingstones in front of the Kogetsu-tei teahouse at the Arisawa residence, Matsue, Shimane Prefecture. Sections of fresh bamboo are placed among the stones to greet special guests invited to the famous Kanden-an tearoom in the building.

236. Steppingstones in garden of Ten'en-an at the temple Nanzen-ji, in Kyoto. The rectilinear border stones and the diagonally set square stones suggest elements of Zen Buddhism.

237. Upward-surging flight of stone steps beyond main gate at the chief Jōdo-sect temple, Chion-in, in Kyoto.

238. Stone lantern of Byōdō-in type. It stands in front of the celebrated tenth-century Phoenix Hall at the Byōdō-in temple, in Uji (near Kyoto). Only the lotus-patterned base is original. The light chamber was added during the Muromachi period. It is thought that this part of the lantern was originally of gold.

239. Pagoda-shaped stone lantern at the famous Nikko shrine Tōshō-gū, dedicated to Tokugawa Ieyasu (1542–1616), the first Tokugawa shōgun. The lantern, which stands next to the ritual basin, was donated to the shrine by the feudal lord Nagai Naomasa (1587–1668), participant in some of the great battles fought by Ieyasu.

240. Stone lantern in front of shrine offices of the Iwashimizu Hachiman-gū, in Kyoto. A classical example of its period, it dates from 1295 and is designated as an important cultural property.

241. Footprints of the Buddha in stone, Kōzan-ji, Kyoto. It is said that before his death and his entry into Nirvana the historical Buddha, Sakyamuni (in Japanese, Shaka), stood on a stone and left his footprints upon it. For centuries afterward, before statues of the Buddha were produced, footprints like these were made and revered. The tradition passed from India into China and thence to Japan, where many copies of the footprints were made. The ones shown here are not old, but the line cutting of the Buddhist symbols on them is quite lovely.

242. Buddhist memorial stupas in ancient style. They stand in the Kaizanchō of the temple Kōzan-ji, in Kyoto. Although such stupas are usually of stone, they may be of wood or metal.

243. Five-level stupa, Kōzan-ji, Kyoto. The five levels, each labeled in Sanskrit, symbolize (from top to bottom) the sky (a jewel shape), the wind (a bowl shape), fire (a truncated pyramid), water (a sphere), and

earth (a square block). From as early as the Heian period, stupas of this type, usually of stone but sometimes of metal or wood, have been made as votive offerings, memorials, or reliquaries at Buddhist temples.

244–45. Stone foxes at the Fushimi Inari Shrine, Kyoto. There are many interpretations of the mythical relations between foxes and the Japanese harvest gods called Inari. One interpretation says that a magician, under the influence of Esoteric Buddhism and Taoism, came to be possessed by fox spirits, developed the ability to make oracular pronouncements, and gradually evolved into the Inari. The Fushimi Inari Shrine is a celebrated place to pray for worldly wealth. The foxes wear aprons labeled "dedication."

246–47. Stone ramparts of moat at Nijō Castle, Kyoto. These walls are the remains of those that surrounded an area of 400 by 500 meters that was once the residence of Ieyasu, the first Tokugawa shōgun. The unmortared construction attests to the ingenuity of the artisans who built them.

248–49. Stone wall at the Onjō-ji, subtemple of the Mii-dera, in Ōtsu, a town on the shore of Lake Biwa. For centuries this area has been renowned for the skill of its stonemasons, excellent examples of whose work are seen in the processional paths of temples and shrines located there. The delicate workmanship in the wall shown here is evidence of their skill.

250. *Suzuribako,* or writing box, containing water pot (upper left), inkstone (lower left), ink stick, and brush. The box is of Shunkei lacquer, a transparent amber-colored lacquer that reveals the grain of the underlying wood. Boxes of this kind are generally lidded and come in many sizes and shapes. They may be lacquered in any of several colors and decorated with gold lacquer or mother-of-pearl inlay. The inkstone is most often of slate.

251. Left: inkstone (*suzuri*) of Nachi black slate from Wakayama Prefecture. A small amount of water is poured into the deep part of the well. The ink stick is then rubbed up and down the slope and across the flat area to produce the desired consistency of ink for painting or writing. The process must be repeated over and over. Right: *suzuri* of Akama stone, a reddish tuff, from San'yō, in Yamaguchi Prefecture.

252. *Ishiusu,* or stone hand mill. Mills like this one have been used in China for many centuries. They were introduced into Japan in the Asuka period (552–646) and at first were used for grinding pigments for painting and for compounding medicines. Later, around the eighteenth century, stonecutting developed to the extent that mills became a common and important article of daily household equipment for grinding buckwheat flour and soy flour and for preparing *tōfu* (bean curd).

253. Mill for extracting oil from rapeseed, sesame seed, and other seeds. Sesame oil is preferred in Japan for deep frying.

254–55. Go board and stones. The stones are kept in wooden bowls. The white ones are polished clam or squilla shell, while the black ones are of the same Nachi slate used in the inkstone on page 251. There are two versions of the history of this tremendously popular game. One says that it originated in India, passed over to China, and later traveled to Japan. The other insists that it was both invented and refined in China.

256. Stone Buddha hands. Buddhist sculpture in stone is rather rare in Japan, and this crude but arresting arrangement of hands found near the Ozawa Pond, in Kyoto, is highly unusual.

257. Red-lacquer sakè cup inscribed in gold with the character *kotobuki* (or *ju*), meaning felicitations and longevity.

258. Kite decorated with character for dragon. The character is highly suitable, since the dragon has the mystical power to fly freely through the sky, stirring up clouds and causing rain to fall.

259. The character *takara,* or treasure, cut from a number of layers of handmade paper with a small knife and hung on the family Buddhist altar as a New Year's decoration.

260–61. Kabuki actors' nameplates. In December, unfinished wooden boards bearing the names and family crests of Kabuki actors are hung in front of the Minamiza (*za* = theater), in Kyoto. The style of writing is characteristic of the Kabuki, and the custom is not limited to Kyoto.

262–63. Traditional fire-brigade jackets. In the Edo period, fires were so frequent in the capital city that they were called the "flowers of Edo," and fire brigades vied with one other in daring acts of bravery. Members of the brigades wore jackets with designs and numbers indicating the individual groups or teams—for example, Number One (page 262) and Number Two (page 263). Jackets like these are still worn in the annual firemen's show held in Tokyo early in January.

264. Feudal-period gold and silver coins, each with its denomination and the name of the issuing authority. The large gold *ōban* at lower right was minted in 1588 under the military dictator Toyotomi Hideyoshi. In 1600, immediately after the victory that gave Tokugawa Ieyasu dominion over most of Japan, a gold guild was established and, in the following year, a silver guild. Above the Hideyoshi *ōban* is an Edo-period *ōban;* at upper left, two Edo-period *koban,* or smaller gold coins; and, at lower left, two silver coins called *chōgin,* also of the Edo period. The *koban* was in everyday use, but etiquette dictated that the *ōban* be used on special occasions.

265. Stone engraved with characters for *gejō,* which means dismounting or alighting. Stones like this one were set up at the gates of certain shrines and temples into whose precincts it was forbidden to ride, either on a horse or in a vehicle. The example shown here is at the temple Daigo-ji, in Kyoto.

266–67. Direction signs indicating south (page 266) and north (page 267), Higashiyama Highway, Kyoto. In the past, signs of wood or stone stood along the main highways to indicate distances and directions for the convenience of Japan's almost exclusively pedestrian traffic.

268. *Senjafuda,* or memorial paper talismans, with names of people and organizations. During the Edo period, as Buddhism became the religion of increasing numbers of ordinary people, there developed the custom of pasting colorful and often quite beautiful *senjafuda* on the ceilings of temples and their main gates. The custom reflects the stylishness—even the raciness—of the urbanites in Edo times.

269. Detail of *fusuma* (sliding partition) with design of stylized peonies and sixteen-petaled chrysanthemums—both imperial emblems. The elaborately worked circular door pulls are also decorated with chrysanthemums. Hokke-ji, Nara.

270–71. Golden imperial chrysanthemums on doors of Gate of the Imperial Messenger, Nishi Hongan-ji, Kyoto. The gate, brought to this temple from Hideyoshi's Fushimi Castle, illustrates the Momoyama-period (1568–1603) flair for exuberance in architecture and decoration. Designed in the Chinese style, it has a gable-end roof of cypress-bark shingles, Chinese gables at front and back, and elaborate carvings. The chrysanthemum crests stand out with dramatic effect.

272–73. *Noren* door curtains in the Gion geisha district, Kyoto. Originally, *noren* were hung at front doors merely to block the sunlight and the lines of vision of passersby. Today they have become symbols of the establishments they grace, serving as advertisements and carrying either the name of the place or some indication of its line of business. The *noren* shown here bear the names of two well-known Gion establishments.

274–75. *Noren* in the Gion district, Kyoto (page 274; see also pages 272–73), and in the city of Takayama, Gifu Prefecture (page 275). The crest on the Gion *noren* represents an old-style well curb. The one on the beautifully woven and dyed Takayama *noren* is *tsuta,* or ivy.

276. Top: short *noren* for household use in interior doorways, where its purpose is largely decorative. This one has the traditional pine, bamboo, and plum blossom in stylized form. Bottom: *furoshiki* wrapping cloth from Izumo, Shimane Prefecture. In this part of Japan it is customary to use *furoshiki* like the one shown here for weddings. In the center is the groom's family crest; in the four corners, the auspicious symbols of the tortoise, the crane, the money-bringing mallet of Daikoku, and the good-fortune scroll.

277. *Fukusa,* or ceremonial cloth, of the kind used under gifts or as a cover for a container. The upper surface is usually of satin, crepe, brocade, or other rich cloth with a family crest embroidered or dyed in the center. The underside is of plain crepe or other silk, and tassels are attached to the corners.

278. Top: shells for shell-matching game. In the Heian period, aristocratic ladies amused themselves

with a game in which scenes from the famous novel *The Tale of Genji* were painted in bright colors and gold on the insides of seashells. Each of 180 pictures was duplicated to make a complete set of 360 shells. Half of these were arranged before the players, while the other 180 were dealt out. The point was to match the dealt shells with the ones set out. Those shown here are modern reproductions in ancient style. Bottom: one-hundred-poets cards. The deck consists of cards with one *tanka* poem each by a group of one hundred famous poets from ancient times down to the Kamakura period. The cards are spread out regularly on the tatami floor, and while one person reads the poems aloud, one by one, the players try to identify the card bearing the poem being read and to snap it up in their hands as quickly as possible. This game is particularly popular at New Year's. The cards shown here are from the temple Hokke-ji, in Nara.

279. *Hanafuda,* or flower cards. Playing cards, introduced into Japan by Europeans in the 1570s, became the basis on which the Japanese developed this distinctive set of cards in suits represented by stylized graphic versions of flowers and animals associated with the twelve months and the four seasons of the year. There are 48 cards in the deck.

280. The hollyhock crest of the house of Tokugawa, shogunate rulers of Japan from 1603 to 1868. This gilded crest is on the light chamber of a lantern at the Tōshō-gū, the shrine in Nikko dedicated to Tokugawa Ieyasu, founder of the shogunate.

The "weathermark" identifies this book as a production of John Weatherhill, Inc., publishers of fine books on Asia and the Pacific. Supervising editor: Ralph Friedrich. Book design and typography: Miriam F. Yamaguchi. Layout of illustrations: Yoshio Hayakawa. Production supervisor: Mitsuo Okado. Composition and printing of the text: Komiyama, Tokyo. Engraving and printing of the plates, in four-color offset and monochrome gravure: Nissha, Kyoto. Binding: Dainihon Bookbinding, Kyoto. The typeface used is Garamond.